Experiments in
NUCLEAR
SCIENCE

Experiments in
NUCLEAR
SCIENCE

Sidney A. Katz ✦ Jeff C. Bryan

CRC Press
Taylor & Francis Group
Boca Raton London New York

CRC Press is an imprint of the
Taylor & Francis Group, an **informa** business

CRC Press
Taylor & Francis Group
6000 Broken Sound Parkway NW, Suite 300
Boca Raton, FL 33487-2742

© 2011 by Taylor and Francis Group, LLC
CRC Press is an imprint of Taylor & Francis Group, an Informa business

No claim to original U.S. Government works

10 9 8 7 6 5 4 3 2 1

International Standard Book Number: 978-1-4398-3481-7 (Paperback)

Library of Congress Cataloging-in-Publication Data

Katz, Sidney A., 1935-
   Experiments in nuclear science / Sidney A. Katz, Jeff C. Bryan.
     p. cm.
   Includes bibliographical references and index.
   ISBN 978-1-4398-3481-7 (pbk. : alk. paper)
   1. Nuclear physics--Experiments. 2. Nuclear chemistry--Experiments. I. Bryan, Jeff
C. II. Title.

QC786.75.K38 2011
539.7078--dc22
                                                   2010023537

Visit the Taylor & Francis Web site at
http://www.taylorandfrancis.com

and the CRC Press Web site at
http://www.crcpress.com

Contents

Preface

This manual was prepared to complement didactic works on radiation biology and radiochemistry as well as those on the other aspects of nuclear science, particularly *Introduction to Nuclear Science* by Jeff C. Bryan, with some relatively uncomplicated laboratory exercises that demonstrate the fundamental principles presented in the textbooks. These experiments were developed for an undergraduate course entitled Radioisotope Methodology, taught with frequent modification for three decades at Rutgers University in Camden. Experiments from Nuclear Chemistry, taught at the University of Wisconsin in La Crosse, have been included to provide both a broader scope and a wider choice. These laboratory exercises have been performed successfully by approximately 500 students under the close supervision of a senior professor.

The exercises contained in this manual are intended to provide opportunities for hands-on experiences with the fundamental principles concerning the origins and properties of nuclear radiations, their detection and measurement, and their applications in basic and applied research, in diagnostic and therapeutic medicine, and in technology and engineering. While the instrumentation for making the measurements on these fundamental aspects of nuclear science has undergone significant changes since the middle of the last century, the fundamental principles of nuclear science have remained virtually unchanged. The instrumentation will, in all probability, continue to change; the fundamental principles, most likely, will not. For this reason, each experiment is introduced with a short review of the underlying theory, while detailed descriptions of the instrumentation are avoided.

With the possible exception of a PuBe neutron source and the liquid scintillation counter, the exercises described in the following pages make use of off-the-shelf equipment and supplies. While a dedicated laboratory is preferred, these experiments can be performed in a conventional chemistry laboratory. Good housekeeping practices and rigorous safety protocols are just as important when working with radioactive materials as they are for all laboratory activities.

Introduction

Nuclear science is a broad discipline dealing with the origins and properties of nuclear radiations, their detection and measurement, and their applications in basic and applied research, in diagnostic and therapeutic medicine, and in technology and engineering. Nuclear science began with Röntgen's (1895) discovery of x-rays and Becquerel's (1896) discovery of radioactivity. Both of these discoveries were made by experiment. The detection and measurement of nuclear radiations preceded the development of nuclear models to explain the phenomenon. The experiments described in this manual are fundamental to understanding the properties of nuclear radiations and their detection and measurement.

Nuclear radiations are the manifestations of energy associated with atomic or nuclear changes. The Rutherford-Bohr model of the atom is adequate for describing most of these changes, such as the increase or decrease in atomic number associated with negatron or positron emission, the decrease in atomic number and mass number associated with alpha emission, x-ray emission associated with electron capture, and Auger electron emission.

The Rutherford-Bohr or planetary model describes the atoms as a central nucleus containing protons and neutrons and orbiting electrons traveling in fairly well defined pathways around the nucleus. In the neutral atom, the number of electrons equals the number of protons.

The atom shown in Figure 1 represents carbon. The nucleus of the carbon atom contains six protons. Six electrons orbit the nucleus of the carbon atom. Each element has a unique name and a unique one- or two-letter symbol. The element carbon has been given the symbol C. The symbol for the element calcium is Ca, and the symbol for the element cadmium is Cd. Each element is placed in the periodic table (Figure 2) according to the number of protons in its nucleus. The number of protons in the nucleus of the atom is the atomic number of the element. The atomic number of carbon is 6 because its nucleus contains six protons. Currently, the periodic table contains some 115 elements.

The nucleus of the atom shown in Figure 1 also contains 6 neutrons. With a single exception, the atoms of all 115 elements contain both protons

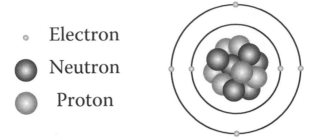

Figure 1 Rutherford–Bohr or planetary model of the carbon atom showing electrons in white, protons in gray, and neutrons in black.

and neutrons. Protons and neutrons are designated as nucleons to indicate they are found in the nucleus of the atom. An assembly of protons and neutrons forming a nucleus is called a nuclide.

At present, some 115 elements are included in the periodic table. Of these, 90 or so are found in nature. These 90 elements exist as about 340 nuclides because most elements, 66 of the 90, exist in nature in more than one isotopic form. For example, the nucleus of the carbon atom shown in Figure 1 contains six protons and six neutrons. It is assigned atomic number 6 because it contains 6 protons. It is assigned mass number 12 because it contains 12 nucleons. A second kind of carbon the nucleus of which contains six protons and seven neutrons also exists in nature. It too is carbon because its nucleus contains six protons. However, its mass number is 13 because its nucleus contains 13 nucleons. To differentiate between them,

1																	2
H																	**He**
3	4											5	6	7	8	9	10
Li	**Be**											**B**	**C**	**N**	**O**	**F**	**Ne**
11	12											13	14	15	16	17	18
Na	**Mg**											**Al**	**Si**	**P**	**S**	**Cl**	**Ar**
19	20	21	22	23	24	25	26	27	28	29	30	31	32	33	34	35	36
K	**Ca**	**Sc**	**Ti**	**V**	**Cr**	**Mn**	**Fe**	**Co**	**Ni**	**Cu**	**Zn**	**Ga**	**Ge**	**As**	**Se**	**Br**	**Kr**
37	38	39	40	41	42	43	44	45	46	47	48	49	50	51	52	53	54
Rb	**Sr**	**Y**	**Zr**	**Nb**	**Mo**	**Tc**	**Ru**	**Rh**	**Pd**	**Ag**	**Cd**	**In**	**Sn**	**Sb**	**Te**	**I**	**Xe**
55	56	57	72	73	74	75	76	77	78	79	80	81	82	83	84	85	86
Cs	**Ba**	**La**	**Hf**	**Ta**	**W**	**Re**	**Os**	**Ir**	**Pt**	**Au**	**Hg**	**Tl**	**Pb**	**Bi**	**Po**	**At**	**Rn**
87	88	89	104	105	106	107	108	109	110	111	112		114		116		118
Fr	**Ra**	**Ac**	**Rf**	**Db**	**Sg**	**Bh**	**Hs**	**Mt**	**Ds**	**Rg**	**Uub**		**Uuq**		**Uuh**		**Uuo**

58	59	60	61	62	63	64	65	66	67	68	69	70	71
Ce	**Pr**	**Nd**	**Pm**	**Sm**	**Eu**	**Gd**	**Tb**	**Dy**	**Ho**	**Er**	**Tm**	**Yb**	**Lu**
90	91	92	93	94	95	96	97	98	99	100	101	102	103
Th	**Pa**	**U**	**Np**	**Pu**	**Am**	**Cm**	**Bk**	**Cf**	**Es**	**Fm**	**Md**	**No**	**Lr**

Figure 2 Periodic table of the elements.

the first is $^{12}_6C$ and the second is $^{13}_6C$. $^{12}_6C$ and $^{13}_6C$ are two isotopes of carbon. Both occur in nature in a ratio of approximately 100 to 1.

Two additional isotopes of carbon are $^{11}_6C$ and $^{14}_6C$. Both are radioactive. These nuclides are unstable. Both spontaneously undergo transitions to more stable arrangements. These transitions are accompanied by the liberation of energy. The latter isotope is formed in the earth's atmosphere by the action of cosmic neutrons on atmospheric nitrogen, $^{14}_7N + {}^1_0n \rightarrow {}^{14}_6C + {}^1_1p$. A steady state is established between the rate of $^{14}_6C$ formation and the rate of $^{14}_6C$ decay. The $^{14}_6C$ equilibrates with natural carbon, the stable isotopes $^{12}_6C$ and $^{13}_6C$. Consequently, all organic life contains 0.255 ± 0.0017 Bq $^{14}_6C$/g carbon = 15.3 ± 0.1 dpm $^{14}_6C$/g carbon (Libby, 1955). Upon death, the equilibrium between $^{14}_6C$ and natural carbon is destroyed. Dead organic matter loses $^{14}_6C$ by radioactive decay. The longer it has been dead, the less radioactive it becomes. This is the basis for carbon dating.

Of the 340 nuclides occurring in nature, 265 are stable. The remaining 75 nuclides are radioactive. In nature, there are 10 radioactive elements (Tc, Po, At, Rn, Fr, Ra, Ac, Th, Pa, and U), each of which exists as several radioactive isotopes. In addition, there are 13 radioisotopes of the stable elements (^3H, ^{14}C, ^{40}K, ^{210}Pb, etc.). *In toto*, more than 2,500 nuclides are known. Most are radioactive, and many do not occur in nature. Those that do not exist in nature are formed as a consequence of fission in nuclear reactors, or carefully produced by deliberate nuclear reaction in research reactors and cyclotrons.

Each nuclide can be described by a unique symbol, $^A_ZE^*$, where A is the mass number, the number of nucleons; Z is the atomic number, the number of protons; E is the chemical symbol; and * is the energy. (Frequently, the subscript, Z, is omitted because the chemical symbol identifies the chemical element, and hence its atomic number.) For example, ^{99}Tc has a half-life of 2.1×10^5 years, while ^{99m}Tc has a half-life of 6.01 hours. ^{99}Tc and ^{99m}Tc are nuclear isomers. Both are ^{99}Tc, but ^{99m}Tc is in a higher or metastable energy state. ^{60}Co and ^{60m}Co are another example of nuclear isomers. Their respective half-lives are 5.25 years and 10.5 minutes. The concept of half-life is discussed in Experiment 11.

The nuclear parameters associated with stability have been described by Bryan (2009), Choppin and Rydberg (1980), Ehmann and Vance (1991), Liester (2001), and Loveland et al. (2006). These parameters include the neutron-to-proton ratio, nucleon pairing, and binding energy.

The neutron-to-proton ratio is frequently described by the band of stability shown in Figure 3. For the lighter elements, $Z < 20$, nuclear stability is associated with a neutron-to-proton ratio of 1 to 1. As the atomic number increases, more than one neutron per proton is associated with nuclear stability. At $Z = 80$, the ratio for stability is 3 to 2.

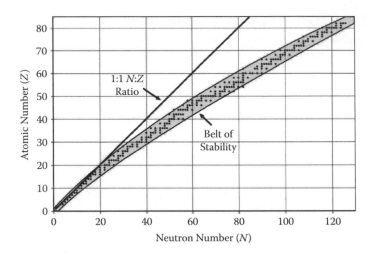

Figure 3 Band of nuclear stability showing the stable nuclides as the dots.

The idea of nucleon pairing is based on the observations listed in Table 1. This table shows the largest majority of the stable nuclides have even numbers of protons, or even Z, and even numbers of neutrons, or even N; the next largest majorities have either even Z and odd N or odd Z and even N; and very few stable nuclides have odd Z and odd N. The pairing of protons with protons or neutrons with neutrons enhances the stability of the nucleus.

Enhanced nuclear stability was observed also when Z = 2, 8, 20, 28, 50, and 82, or when N = 2, 8, 20, 28, 50, 82, and 126. These values for the numbers of protons and numbers of neutrons were (and still are) referred to as magic numbers. This led to the idea of filled nuclear shells for protons and neutrons, respectively. The development of a nuclear shell model for the nucleus paralleled that for the electron shells in the Rutherford-Bohr model of the atom described above. The nuclear shell model is based on empirical evidence and theoretical considerations (Goeppert-Mayer, 1948, 1950a, 1950b). Support for the nuclear shell model comes from several observations:

Table 1 Distributions of Nucleons in the Stable Nuclides

Number of stable nuclides	Z	N
158	Even	Even
53	Even	Odd
50	Odd	Even
4	Odd	Odd

- The decay series for the three naturally occurring nuclides, ^{232}Th, ^{235}U, and ^{238}U, all end with stable isotopes of lead, Z = 82.
- The most stable isotopes of the heaviest nuclides, lead, and bismuth have 126 neutrons.
- Calcium, Z = 20, exists as six stable isotopes, while potassium, Z = 19, and scandium, Z = 21, exist as two and one stable isotope, respectively.
- ^{40}Ca is "double magic," with 20 protons and 20 neutrons. The isotopic abundance of ^{40}Ca is 96.941%.
- Similarly, the double magic ^{208}Pb with 82 protons and 126 neutrons is the most abundant of the stable lead isotopes.
- Nickel, Z = 28, exists as five stable isotopes while cobalt, Z = 27, and copper, Z = 29, exist as one and two stable isotopes, respectively.
- Similarly, tin Z = 50, exists as 10 stable isotopes while indium, Z = 49, and antimony, Z = 51, each exists as two stable isotopes.
- Nuclides one neutron short of a magic number have high probabilities, σ values, for neutron capture.
- ^{87}Kr and ^{137}Xe, both of which are one neutron above a magic number, are delayed neutron emitters.

Nuclear binding energy is the energy equivalent of the mass difference between the nuclide and its constituent nucleons. For example, the binding energy of the $^{4}_{2}$He nucleus is 28.30 MeV. The mass of the $^{4}_{2}$He nucleus is 4.00150618 u, and it consists of two protons and two neutrons. The rest mass of a proton is 1.00727647 u, and that of a neutron is 1.00866442 u. The mass difference is [4.00150618 − (2 × 1.00727647 + 2 × 1.00866442)] = 0.0303765 u. The energy equivalent of this mass difference is [0.0303765 u (9.315 × 10^3 MeV/u)] = 28.30 MeV. Comparisons between nuclides are made on the basis of binding energy per nucleon. The greater the binding energy, the more stable is the nuclide. The binding energy per nucleon for $^{4}_{2}$He is 7.075 MeV per nucleon, i.e., 28.30 ÷ 4. Similarly, the binding energies per nucleon for $^{210}_{84}$Po and $^{206}_{82}$Pb are 7.833 and 7.874 MeV per nucleon, respectively. Question: Confirm these values by calculation and comment on which nuclide, $^{210}_{84}$Po or $^{206}_{82}$Pb, is more stable.

The transition from an unstable to a stable nuclear configuration results in the liberations of energy. The rate of transition and the manifestation of the energy liberated are characteristics of the radioactive nuclide. The rate is described by the half-life of the radioactive nuclide, and the energy manifestations are nuclear radiations designated as alpha (α), beta (β), or gamma (γ). Half-life is discussed in Experiment 11.

The $^{210}_{84}$Po mentioned above has a lower binding energy per nucleon than does the $^{206}_{82}$Pb. The respective nuclear binding energies are 1,645 and 1,622 MeV. They differ by 23 MeV and four nucleons, two protons, and two neutrons. The removal of two protons and two neutrons from $^{210}_{84}$Po requires 23 MeV. As seen above, the formation of $^{4}_{2}$He from two protons and two neutrons liberates 28.30 MeV. Consequently, $^{210}_{84}$Po → $^{206}_{82}$Pb + α is energetically feasible. The difference between 28 MeV and 23 MeV is 5 MeV. This is the approximate kinetic energy of the α emitted in the transition from $^{210}_{84}$Po to $^{206}_{82}$Pb. The logarithm of the half-life of the α-emitting nuclide is inversely proportional to the reciprocal of the square root of the energy of the α, E_α (Geiger and Nuttall, 1911):

$$\ln t_{1/2} = \frac{k_1 Z}{\sqrt{E_\alpha}} + k_2$$

The alpha (α) as well as the beta (β) and gamma (γ) radiations are manifestations of the energy liberated in the course of radioactive decay. These nuclear radiations dissipate energy as they transverse matter. The magnitude of the energy is a characteristic of the transition from unstable nuclide to stable nuclide. The radioactive decay of unstable $^{210}_{84}$Po to stable $^{206}_{82}$Pb involves the emission of α radiation, the energy of which is 5.408 MeV.

The $^{14}_{6}$C described above decays with a half-life of 5,730 years by emitting β radiation that has a maximum energy, E_{max}, of 0.156 MeV. The half-life of the $^{11}_{6}$C described above is 20.3 minutes, and the maximum energy of the β radiation is 0.970 MeV. The β radiation associated with the decay of $^{14}_{6}$C is the negatron, β^-, while that associated with the decay of the $^{11}_{6}$C is its "antiparticle," the positron, β^+. Each unstable nuclide undergoes radioactive decay at a characteristic rate with the emission of a characteristic radiation. The decay scheme for $^{198}_{79}$Au, shown in Figure 4, is more complicated. $^{198}_{79}$Au is a β/γ emitter while $^{14}_{6}$C is a pure β emitter. The information contained in Figure 4 shows the radioactive decay of $^{198}_{79}$Au proceeds at a rate corresponding to a half-life of 2.7 days. Most (98.99%) of the decays involve emission of a β^- having a maximum energy of 0.960 MeV and a γ with an energy of 0.412 MeV. Knowledge of the decay scheme is needed for the efficient detection and measurement of radioactive materials. Information on these decay schemes can be found in *Table of Isotopes* (Lederer et al., 1967) and elsewhere (Friedlander et al., 1981).

Rutherford identified α radiation as identical to high-energy $^4_2$He^{+2} ions. The β radiation is identical to high-energy electrons, e^{-1}, or high-energy positrons, e^{+1}. When the positron interacts with its antiparticle, the electron, both particles are annihilated. Energy equivalent to the mass of both particles is liberated in accord with the Einstein equation of 1905: $\Delta E = \Delta mc^2$. The energy liberated is 1.022 MeV, the equivalent of two electron masses, i.e., $2 \times 5.486 \times 10^{-4}$ u $\times$ 931.5 MeV/u = 1.022 MeV. Most frequently, this annihilation radiation appears as two 0.511 MeV photons directed 180° apart. These photons associated with annihilation radiation, like the gamma radiation associated with radioactive decay, can be described as

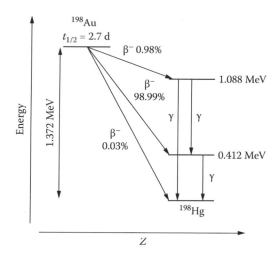

Figure 4 Decay scheme for ¹⁹⁸Au.

energy moving through space. The characteristics of the nuclear radiations are summarized in Table 2.

Alpha radiations from the radioactive decay of a given nuclide are monoenergetic, while beta radiations show a spectrum of energies ranging from zero up to a characteristic maximum, E_{max}. The 0.156 MeV β associated with the decay of ¹⁴C is the maximum energy of the radiations. Most have energies of approximately 0.05 MeV. The average energy is approximately one-third of the maximum energy. Gamma radiations associated with the radioactive decay of a given nuclide are monoenergetic. Neutron emission is rare. The differences in the characteristics of α, β, and γ radiations give rise to differences in their properties and, consequently, differences in techniques for the efficient detection and measurement of radioactive materials.

Alpha decay is most commonly observed in nuclides with atomic numbers greater than 83. When a nuclide decays by alpha emission, the product will be two atomic numbers and four mass numbers less, i.e.,

Table 2 Characteristics of Nuclear Radiations

Radiation	Name	Charge	Rest mass[a]
α	Alpha	+2	4.00150618
β	Beta	±1	0.00054858
γ	Gamma	0	0
n	Neutron	0	1.00866492
p	Proton	+1	1.00727647

[a] Mass is relativistic.

$^{226}_{88}Ra \rightarrow {}^{222}_{86}Rn + \alpha$. Remember, the α is a ^4_2He nucleus. Note: In a nuclear equation, the sum of the mass numbers of the reactants will equal the sum of the mass numbers of the products, and the sum of the atomic numbers of the reactants will equal the sum of the atomic numbers of the products. The energy of the alpha radiation from the decay of $^{226}_{88}Ra$ is 4.781 MeV. All of the alpha particles emitted from the $^{226}_{88}Ra$ have this energy. They are monoenergetic. $^{222}_{86}Rn$ also decays by alpha emission. The product of this decay must have Z = 86 − 2 = 84 and A = 222 − 4 = 218. Question: Why must the product be an isotope of polonium? The energy of the alpha radiation from the decay of $^{222}_{86}Rn$ is 5.587 MeV.

The half-lives of $^{226}_{88}Ra$, $^{222}_{86}Rn$, and $^{210}_{84}Po$ are 1,600 years, 3.80 days, and 138 days, respectively, and the corresponding E_α values are 4.781, 5.587, and 5.408 MeV. Question: Do these α-emitting radionuclides follow the Geiger-Nuttall relationship cited above?

Negatron emission is a likely mode of decay for nuclides with a high neutron-to-proton ratio. These are the neutron-rich nuclides located below the band of stability shown in Figure 3.

The nuclide $^{32}_{15}P$ is neutron rich. It consists of 15 protons and [32 − 15 =] 17 neutrons. The loss of a neutron with the gain of a proton accompanied by emission of a negatron results in a more stable composition: $n \rightarrow p^{+1} + \beta^{-1}$. The nuclide resulting from such a change will have 16 protons and 16 neutrons, and it will be $^{32}_{16}S$. Unstable $^{32}_{15}P$ with an odd-odd neutron-proton arrangement and a neutron-to-proton ratio of >1 to 1 can achieve a more stable even-even neutron-proton arrangement, and a 1-to-1 neutron-to-proton ratio by negatron emission.

The energy associated with this transition is 1.71 MeV based on the difference between the masses of $^{32}_{16}P$ and $^{32}_{15}S$, 31.97391 and 31.97207 u, respectively. This energy is partitioned between the negatron and the antineutrino. Sometimes, the entire 1.71 MeV is imparted to the negatron and none to the antineutrino. Hence, the maximum energy of a negatron from the decay of $^{32}_{16}P$, or its E_{max}, is 1.71 MeV. This happens only rarely. More often, the partition is closer to 1 to 2, and the average energy of a negatron from the decay of ^{32}P is approximately 0.5 to 0.6 MeV. Correspondingly, the average energy of the antineutrino from the decay of $^{32}_{16}P$ is approximately 1.1 to 1.2 MeV.

The nuclide $^{24}_{11}Na$, like $^{32}_{15}P$, is odd-odd and neutron rich. It too decays by negatron emission with an E_{max} of 1.39 MeV, but the transition from $^{24}_{11}Na$ to $^{24}_{12}Mg$ leaves residual energy in the product nuclide. This residual energy is emitted as the two gamma rays with energies of 1.38 and 2.76 Mev, respectively, as shown in Figure 5. The nuclide $^{24}_{11}Na$ is an example of a β/γ emitter, while the nuclide $^{32}_{15}P$ and the $^{14}_{12}C$ cited earlier are examples of pure beta emitters.

Decay of the proton-rich nuclides lying above the band of stability in Figure 3 is more complicated. The decay scheme for $^{22}_{11}Na$, an odd-odd, proton-rich nuclide, is shown in Figure 6. $^{22}_{11}Na$ decays to $^{22}_{10}Ne$ in two

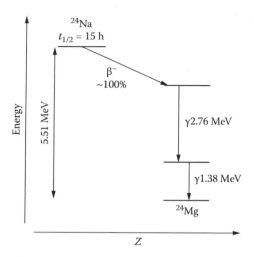

Figure 5 Decay scheme for ^{24}Na.

ways: 89.8% of the time the decay is by positron emission, and 10.2% of the time the decay is by electron capture. E_{max} for the positron is 0.545 MeV. Both positron emission and electron capture are accompanied by the emission of 1.275 MeV gamma radiation. When the energy change associated with the radioactive decay is in excess of 1.02 MeV, positron emission and electron capture can be competing modes of decay. In the case of $^{22}_{11}$Na, the total energy associated with the decay is 1.02 MeV + 0.545 MeV + 1.275MeV = 2.85 MeV.

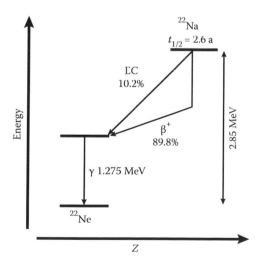

Figure 6 Decay scheme for ^{22}Na.

As mentioned above, the positron is annihilated in a collision with its antiparticle, the negatron, and the energy equivalent of the two electron masses appears as two 0.51 MeV photons. This annihilation radiation is characteristic of all positron emitters.

Electron capture is often accompanied by emission of a characteristic gamma photon. In addition, electron capture, which most frequently involves a K shell electron, is accompanied by x-ray emission characteristic of the product nuclide as outer shell electrons assume lower energy levels closer to the nucleus.

Sometimes, a fraction of these x-rays interacts with orbital electrons. The energy of these x-rays is absorbed by the electrons, and the electrons are ejected from their orbits. They differ from beta radiations in that they are monoenergetic, having energies equal to that of the x-ray less their binding energies. They are referred to as Auger electrons. Even though this phenomenon was first observed by Meitner (1923), the name honors Auger (1925) for the work he reported two years later. A similar phenomenon occurs when gamma radiations are absorbed by orbital electrons. Like the Auger electrons, these conversion electrons are monoenergeric.

Radioactive nuclides undergo decay by characteristic decay schemes. These decay schemes can include alpha emission, negatron emission, negatron emission with gamma emission, conversion electron emission, positron emission, annihilation radiation, electron capture, electron capture with gamma emission, and Auger electron emission. It is important to understand the decay scheme of a radionuclide and the radiations it emits in order to efficiently detect and measure it.

Laboratory safety

Nuclear radiations dissipate their energies as they travel through matter. Most often, the energy of the radiation is consumed by interactions with the orbital electrons of the matter transversed. This absorbed energy results in the formations of a positive ion and a negative electron, i.e., an ion pair. When the matter transversed is a living system, the dissipated energy is absorbed by the system and causes damage to the system. The extent of damage depends upon the kind of radiation, the energy of the radiation, the intensity of the radiation, the tissue absorbing the energy, and the duration of the exposure to the radiation as well as individual characteristics, such as age, state of health, etc. While the different types of radiation have differing abilities to damage tissue, all radiations are able to damage tissue. For this reason, all radiation must be considered to be potential hazards.

Tissue damage by radiation shows a dose-response. Exposures to high doses of radiation cause extensive tissue damage. The likelihood of such external exposures from the quantities of radioactive materials encountered in the course of executing the experiments described in this manual is remote. Internal exposures are of much greater concern.

Exposure to nuclear radiation is expressed as the exposure dose. Of more concern is the absorbed dose, the fraction of the exposure dose that is absorbed.

Prior to 2006, exposure dose was measured in röntgens, R, where one röntgen was the quantity of electromagnetic (x or γ) radiation required to liberate one electrostatic unit (esu) of positive and negative charge in 1 cm^3 of dry air at standard atmospheric pressure (STP). One röntgen produces approximately 2.08×10^9 pairs of positive and negative ions in 1 cm^3 of dry air at STP. The röntgen is not an SI unit. In SI units, 1 R $= 2.58 \times 10^{-4}$ C/kg (calculated from 1 esu $= 3.33564 \times 10^{-10}$ C and 1.293 kg/m^3 as the standard atmospheric density of air). The röntgen is a measure of the energy dissipated in air.

Other units for radiation exposure are defined in Part 20, § 20.1004 of Title 10 (NRC, 2008), as follows:

> (a) Definitions. As used in this part, the units of radiation dose are:
>   - *Rad* is the special unit of absorbed dose. One rad is equal to an absorbed dose of 100 ergs/gram or 0.01 joule/kilogram (1 rad = 0.01 gray).
>   - *Gray* (Gy) is the SI unit of absorbed dose. One gray is equal to an absorbed dose of 1 Joule/kilogram (100 rads).
>   - *Rem* is the special unit of any of the quantities expressed as dose equivalent. The dose equivalent in rems is equal to the absorbed dose in rads multiplied by the quality factor (1 rem = 0.01 sievert).
>   - *Sievert* is the SI unit of any of the quantities expressed as dose equivalent. The dose equivalent in sieverts is equal to the absorbed dose in grays multiplied by the quality factor (1 Sv = 100 rems).
> (b) As used in this part, the quality factors for converting absorbed dose to dose equivalent are shown in Table 3.

Exposures to lower doses of radiation result in less tissue damage than do high-dose exposures. Every effort is made to minimize radiation exposures to as low as reasonably achievable (ALARA), and under no conditions exceeding those limits specified in Part 20, § 20.1201 of Title 10 (NRC, 2008). These exposure limits cited in Title 10 are as follows:

Table 3 Definitions of Dose

Type of radiation	Quality factor (Q)	Absorbed dose equal to a unit dose equivalent[a]
X, gamma, or beta radiation	1	1
Alpha particles, multiple charged particles, fission fragments, and heavy particles of unknown charge	20	0.05
Neutrons of unknown energy	10	0.1
High-energy protons	10	0.1

[a] Absorbed dose in rad equal to 1 rem, or the absorbed dose in gray equal to 1 sievert.

(a) The licensee shall control the occupational dose to individual adults, except for planned special exposures under § 20.1206, to the following dose limits.

    (1) An annual limit, which is the more limiting of

        (i) The total effective dose equivalent being equal to 5 rems (0.05 Sv); or

        (ii) The sum of the deep-dose equivalent and the committed dose equivalent to any individual organ or tissue other than the lens of the eye being equal to 50 rems (0.5 Sv).

    (2) The annual limits to the lens of the eye, to the skin of the whole body, and to the skin of the extremities, which are:

        (i) A lens dose equivalent of 15 rems (0.15 Sv), and

        (ii) A shallow-dose equivalent of 50 rems (0.5 Sv) to the skin of the whole body or to the skin of any extremity.

(b) Doses received in excess of the annual limits, including doses received during accidents, emergencies, and planned special exposures, must be subtracted from the limits for planned special exposures that the individual may receive during the current year (see § 20.1206(e)(1)) and during the individual's lifetime (see § 20.1206(e)(2)).

(c) When the external exposure is determined by measurement with an external personal monitoring device, the deep-dose equivalent must be used in place of the effective dose equivalent, unless the effective dose equivalent is determined by a dosimetry method approved by the NRC. The assigned deep-dose equivalent must be for the part of the body receiving the highest exposure. The assigned shallow-dose equivalent must be the dose averaged over the contiguous 10 square centimeters of skin receiving the highest exposure. The deep-dose equivalent, lens-dose equivalent, and shallow-dose equivalent may be assessed from surveys or other radiation measurements for the purpose of demonstrating compliance with the occupational dose limits, if the individual monitoring device was not in the region of highest potential exposure, or the results of individual monitoring are unavailable.

(d) Derived air concentration (DAC) and annual limit on intake (ALI) values are presented in Table 1 of appendix B to part 20 and may be used to determine the individual's dose (see § 20.2106) and to demonstrate compliance with the occupational dose limits.

(e) In addition to the annual dose limits, the licensee shall limit the soluble uranium intake by an individual to 10 milligrams in a week in consideration of chemical toxicity (see footnote 3 of appendix B to part 20).

(f) The licensee shall reduce the dose that an individual may be allowed to receive in the current year by the amount of occupational dose received while employed by any other person (see § 20.2104(e)). [56 FR 23396, May 21, 1991, as amended at 60 FR 20185, April 25, 1995; 63 FR 39482, July 23, 1998; 67 FR 16304, April 5, 2002; 72 FR 68059, December 4, 2007.]

The effects of high-dose, external exposures are summarized in Table 4.

Table 4 Probable Health Effects Resulting from Whole Body Exposure to
Ionizing Radiation

Dose in rems	Immediate health effects	Delayed health effects
>1,000	Immediate death	None
600–1,000	Weakness, nausea, vomiting, and diarrhea followed by apparent improvement; after several days, fever, diarrhea, bloody bowel discharge, hemorrhage of the larynx, trachea, and lungs, vomiting blood, bloody urine	Death in about 10 days; autopsy findings show destruction of bone marrow and other blood-forming tissues, and swelling and degeneration of the epithelial cells of the intestine, genital organs, and endocrine glands
250–800	Nausea, vomiting, diarrhea, epilation, weakness, bloody bowel discharge, nose bleeds, bleeding gums, inflamed pharynx, and stomach; marked destruction of bone marrow, lymph nodes, and spleen causes decrease in blood cells particularly granulocytes and thrombocytes	Atrophy of some endocrine glands; between weeks 3 and 5 postexposure, death is correlated with degree of leukocytopenia; 50% die in this period; survivors experience ophthalmologic disorder, blood dyscrasis, tumors, psychological disturbance
150–250	Nausea and vomiting on first day; diarrhea and skin burns; apparent improvement for about 2 weeks; fetal or embryonic death if pregnant	Symptoms of malaise as indicated above; persons in poor health prior to exposure or those who develop serious infection may not survive; the healthy adult recovers to somewhat normal health in approximately 3 months; he or she may have permanent health damage, develop cancer or benign tumors, and will probably have a shortened life span; genetic and teratogenic effects

(continued)

Table 4 Probable Health Effects Resulting from Whole Body Exposure to
Ionizing Radiation (Continued)

Dose in rems	Immediate health effects	Delayed health effects
50–150	Acute radiation sickness and burns are less severe than at higher exposure doses; spontaneous abortion or stillbirth	Tissue damage effects are less severe; reduction in lymphocytes and neutrophils leaves the individual temporarily very vulnerable to infection; there may be genetic damage to offspring, benign or malignant tumors, premature aging, and shortened life span; genetic and teratogenic effects
0–10	None	Premature aging, mild mutations in offspring, some risk of excess tumors; genetic and teratogenic effects

Factors for reducing exposures from external sources of radiation include time, distance, and shielding. Application of these factors involves:

1. Working quickly and carefully
2. Using forceps or remote manipulators when handling radioactive sources
3. Placing inert materials between the user and the radioactive materials

Internal radiation exposure is much more serious and requires special care to prevent inhalation, ingestion, and dermal absorption of radioactive materials. Strict adherence to the radioisotope laboratory safety rules is mandatory.

Radioisotope laboratory safety rules

1. Outerwear, books, and other personal belongings may not be brought into any laboratory where unsealed radioactive materials are used or stored.
2. Visitors to the laboratory must receive permission and authorization to enter.
3. The use of laboratory coats and laboratory gloves is strongly recommended. Once brought into the working laboratory, these items may not be removed from the laboratory.
4. Eating and drinking, storing and preparing food, and applying cosmetics are prohibited in all laboratories.

5. Direct contact with radioactive materials must be avoided.
6. Physical manipulations involving radioactive materials must be performed in a shallow, paper-lined tray.
7. Pipetting by oral suction is not permitted.
8. Spills and losses of radioactive materials must be reported immediately.
9. Injuries must be reported immediately.
10. Real or imagined exposures to radiation and radioactive materials must be reported immediately.
11. Complete records of receipt, transfer, and disposal of all radioactive materials must be maintained, and radiation exposure must be monitored and recorded.
12. Operations involving the possible release of radioactive gases, vapors, mists, or dusts must be confined to an approved hood or glove box.
13. All radioactive materials must be properly labeled, handled, and stored.
14. All radioactive wastes must be properly labeled and properly managed.
15. At the end of the working period, all glassware and equipment must be decontaminated and monitored before they are returned to general use.
16. At the end of the working period, all bench tops and other working surfaces must be monitored and, if necessary, decontaminated.
17. At the end of the working period, hands as well as other parts of the body likely to become contaminated with radioactive material must be monitored, decontaminated, and monitored again.
18. Adherence to these radioisotope laboratory safety rules is mandatory.

Laboratory organization

Preparation for the laboratory is required. It is expected that each laboratory worker will have reviewed and understood the experimental procedures prior to commencing work.

Laboratory safety rules and all other written and verbal instructions must be observed.

Each workstation should be equipped with a shallow, paper-lined tray containing the items required for the experiment. With the exception of the actual measurement of radioactivity, manipulations involving radioactive materials must be confined to this tray. The transport of radioactive materials from the workstation to the measurement station must be made in an approved container.

Most of the experiments described below can be executed with a scaler-timer and a Geiger-Müller detector. The instrumentation shown in Figure 7 is relatively uncomplicated and fairly robust. However, it does

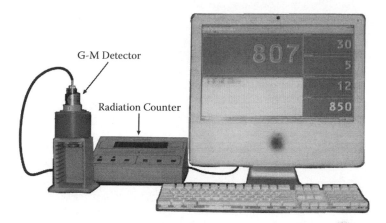

Figure 7 Scaler-timer and Geiger-Müller detector.

not tolerate abuse. Special care must be taken to protect the fragile window of the Geiger-Müller detector. The various functions of the counting software should be understood before they are activated. The common features include an adjustable high-voltage power supply to the detector, a count timer, a start switch, and a display for recording the number of events. Some additional features are the ability to preset the number of counts and runs and export data to graphing software.

The safety rules apply during all laboratory sessions.

chapter one

Experiment 1 – characteristics of Geiger-Müller counters

The objectives of this experiment are to establish the dependence of the Geiger-Müller (G-M) counter response as a function of the potential applied to the anode, to determine the operating potential of the Geiger-Müller counter and measure the efficiency of the Geiger-Müller counter for the detection and measurement of nuclear radiations.

The Geiger-Müller detector described in Figure 8 consists of a cylindrical cathode containing a gas of low ionization potential and a central anode. The anode is often charged several hundred volts positive relative to the cylindrical cathode. The filling gas is frequently argon containing a few percent of a polyatomic hydrocarbon or a diatomic halogen. When ionizing radiation passes into the counter, the energy of the radiation is partially or wholly dissipated by ionizing the filling gas. On average, the energy to form one ion pair in argon, $Ar^+ + e^-$, is 26 eV. The β radiation associated with the decay of ^{14}C, having an average energy of approximately 0.05 MeV, could produce approximately 2,000 ion pairs. The electrons from this primary ionization accelerate toward the anode and acquire sufficiently high energy to initiate secondary, tertiary, and subsequent ionizations. These multiplications increase the number of electrons eventually collected as an avalanche at the anode by a factor on the order of 10^{10}. The multiplication depends, in part, on the shape of the anode and the difference in potential between the anode and the cathode. Consequently, the response of a Geiger-Müller counter has a characteristic dependence on the potential applied to the anode, as shown in Figure 9.

A second characteristic of the Geiger-Müller counter is a period of insensitivity following the electron avalanche caused by the sheath of positive argon ions surrounding the anode. This period of insensitivity, or resolving time, is considered in Experiment 2.

The Geiger plateau shown in Figure 9 has a slight positive slope. Some of the factors responsible for the plateau slope have been elaborated upon by Spatz (1943). Although small, this slope can be partially responsible for the day-to-day variations in the detection efficiency. Corrections for these variations are considered in Experiment 11.

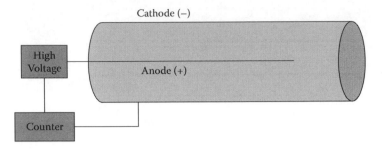

Figure 8 The Geiger-Müller detector.

Procedure

a. Place a radioactive source under a G-M counter.
b. With the high voltage at the lowest setting, make a 1-minute count and record the result.
c. Increase the high voltage by 50 volts, and make another 1-minute count. Record this result.
d. Repeat step (c) several times. Do not exceed the threshold potential by more than 250 volts.
e. After determining the operating voltage, V_O, measure the activities of several calibrated reference sources at this voltage.

Experimental data

Voltage Count Rate	Voltage Count Rate	Voltage Count Rate

Calibrated standard   _____   _____   _____

Observed activity at V_O _____ cpm   _____ cpm   _____ cpm

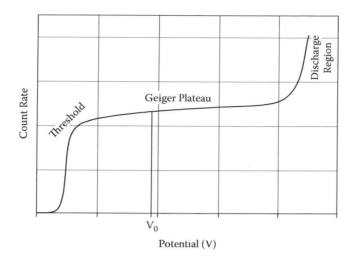

Figure 9 Characteristic voltage response of a Geiger-Müller counter.

Report

a. Using Microsoft Excel™ or similar, plot the count rate, R, against the voltage, V, and determine the operating voltage, V_O, of the Geiger-Müller detector.

b. Determine the efficiencies with which the calibrated reference sources were measured at V_O.

Efficiencies

Source _____ dpm _____ cpm _____ Efficiency_____

Source _____ dpm _____ cpm _____ Efficiency_____

Source _____ dpm _____ cpm _____ Efficiency_____

Source _____ dpm _____ cpm _____ Efficiency_____

c. Explain any differences observed in the efficiencies.

Questions

1. For which radionuclides, ^{14}C, ^{3}H, ^{32}P, ^{99}Tc, ^{90}Y, does the Geiger-Müller detector show the highest and lowest efficiencies? Explain briefly.

2. What activity in cpm is expected from a 0.035 µCi of ^{32}P when it is measured with 5.4% efficiency?

3. Assuming the gas amplification factor is 10^{10} and, on average, the energy to form one ion pair in argon, $Ar^+ + e^-$, is 26 eV, estimate the maximum number of electrons collected at the anode of a Geiger-Müller detector when the β radiation associated with the decay of ^{14}C is absorbed in the gas. Remember, the average energy of this β radiation is 0.050 MeV.

4. Would it be possible to determine the operating voltage if a source emitting a different type of radiation were used? For example, if a beta-emitting source were used in this experiment, would a gamma source give approximately the same result?

5. Why is it a good idea to periodically check the high voltage (HV) plateau for a G-M detector?

6. Is it possible to estimate the energy (E_{max}) of a beta-emitting nuclide using a G-M detector?

7. Based on the data collected, should the user have confidence in the G-M detector?

chapter two

Experiment 2 – resolving time

The objectives of this experiment are to collect data demonstrating resolving time losses and to apply several approaches for approximating the resolving time.

As mentioned in Experiment 1, the Geiger-Müller counter has a period of insensitivity following the electron avalanche caused by the sheath of positive argon ions surrounding the anode. This period of insensitivity, or resolving time, can be evaluated from the data collected by measuring the activities of split sources individually and combined. Sometimes, resolving time is called dead time or paralysis time.

The split-source kit consists of two semicircular sources, S_1 and S_2, and a semicircular blank, B. The sequence for measuring their activities should be S_1 and B, S_1 and S_2, B and S_2, and B alone. The corresponding count rates will be r_1, r_{12}, r_2, and r_B, where r is the observed count rate. Due to resolving time loss, $(r_1 + r_2) > (r_{12} + r_B)$. If there were no resolving time loss, $R_1 + R_2 = R_{12} + R_B$, where R is the true count rate. It follows that during unit time, the ratio of observed count rate to the true count rate will be $R/r = 1/(1- r\tau)$, where τ is the resolving time.

The dead time in milliseconds can be calculated using the Schiff formula (Schiff, 1936) and several of its modifications (shown below), such as those recommended by Preuss (1952), Friedlander et al. (1981), Choppin and Rydberg (1980), and Chase and Rabinowitz (1962), or when the background is high, the hybrid dead time model recommended by Lee and his collaborators (Lee et al., 2004).

$$\tau = (r_1 + r_2 - r_{12}) \div (2\ r_1 r_2)$$

$$\tau = (r_1 + r_2 - r_{12} - r_B) \div (r_{12}^2 - r_1^2 - r_2^2)$$

$$\tau = [(r_1 + r_2 - r_{12}) \div (2\ r_1 r_2)][1 \div 8\ r_{12}\ 2\ r_B\ ][(r_1 + r_2 - r_{12})\ r_{12} \div (r_1\ r_2)]$$

Count rates can be corrected for resolving time loss using the relationship $R = r/(1- r\tau)$.

Alternatively, an empirical correction for a Geiger-Müller counter having a resolving time of 300 μs can be applied by adding 0.5% of the observed count rate for each 1,000 cpm recorded. Lambie (1964) has published a tabulation of corrections for a Geiger-Müller counter having a resolving time of 400 μs, and Carswell (1967) has published graphical corrections for Geiger-Müller counters having resolving times of 300, 400, and 500 μs.

Procedure

Measure the activity of the split sources individually and combined, being certain to maintain constant positioning.

Experimental data

	Trial 1	Trial 2	Trial 3
r_1	_____	_____	_____
$r_{1,2}$	_____	_____	_____
r_2	_____	_____	_____
r_b	_____	_____	_____

Report

 a. Calculate the mean values for r_1, $r_{1,2}$, r_2, and r_b.
 b. From these mean values, calculate the resolving time, τ, using the three equations as well as the 0.5% per 1,000 approximation cited above.

$\tau =$ _____ μs
$\tau =$ _____ μs
$\tau =$ _____ μs
$\tau =$ _____ μs

Questions

 1. Compare the resolving time loss when 5,530 cpm is recorded with a Geiger-Müller detector having a dead time of 335 μs using the 0.5% per 1,000 approximation with that obtained from the expression $R/r = 1/(1- r\tau)$.
 2. The coincidence correction for a particular GM counter is 6.0% at count rates of 10,000 cpm. Calculate the true count of a sample that counts 10,500 cpm on this counter.
 3. Would it have made a significant difference in determining the operating voltage in the previous experiment if the count rate had been 2 or 10 times greater?
 4. Why is it important to maintain constant positioning during this experiment?

5. At a relatively low count rate (r_{12} < 1,000 cpm), a student observes that ($r_1 + r_2$) < ($r_{12} + r_B$). Can you explain this apparent anomaly?
6. Why are three trials performed in this experiment?
7. Comment on the consistency (or lack thereof) in the four values of τ determined in this experiment.

chapter three

Experiment 3 – background corrections

The objectives of this experiment are to demonstrate the presence of background radiation and the utilization of massive shielding to reduce the background.

The result of a laboratory measurement on a radioactive sample will include both the radiations from the sample and the background radiation. The background radiation has several components. Among them are cosmic radiations, radiations from the naturally occurring radionuclides in the geosphere and atmosphere, as well as radiations from the $^{40}_{19}K$, $^{14}_{6}C$, etc., present in living tissue.

A second source of apparent background activity for scintillation counters is due to thermionic emission at the photocathode of the photomultiplier tube. These false counts come from the random ejection and subsequent multiplication of electrons from the photocathode of a photomultiplier tube. The operation of the photomultiplier tube is described in Experiment 18.

It is customary to correct a gross count by subtracting a separately measured background count to obtain the net count of the sample. This approach is acceptable when the gross count is large and the background count is small. When the difference between the gross count and the background count is small, the error in the net count is large. For example, the uncertainty in the net count rate is ±8.4 cpm when the gross count collected in 1 minute is 40 and the background count collected in 1 minute is 30, i.e.,

$$(40 \pm \sqrt{40}) - (30 \pm \sqrt{30}) = 10 \pm \sqrt{(6.3^2 + 5.5^2)} = 10 \pm 8.4 \text{ cpm}$$

The uncertainty associated with a single count is the square root of that count (Bryan, 2009, p. 33; Loveland et al., 2006, p. 573).

The uncertainty in the net count rate can be reduced by increasing the duration of time from 1 minute to 10 minutes, i.e.,

$$(400 \pm \sqrt{400}) - (300 \pm \sqrt{300}) = 100 \pm \sqrt{(20^2 + 17.3^2)} = 100 \pm 26 \text{ in 10 minutes,}$$
or 10 ± 2.6 cpm

The relative error is reduced from 84% to 26% by increasing the count time from 1 minute to 10 minutes.

In addition, the error in the net count rate can be reduced by elaborate electronic circuitry and by shielding the detector from external radiation.

Procedure

Measure the background count rates for shielded and unshielded Geiger-Müller counters.

Experimental data

	Trial 1	Trial 2	Trial 3
Count rate, unshielded, cpm	_____	_____	_____
Count rate, shielded, cpm	_____	_____	_____

Report

a. Comment on the effects of shielding on background count rates.
b. Calculate the percent reduction in background count rate by using shielding.

Background reduction, %   _____   _____   _____

Questions

1. What is the most likely source of background radiation?
2. What type of radiation is likely most commonly being measured in this experiment (α, β, or γ)?
3. How could the percent reductions in background count rate be lowered even more?
4. In the previous experiment, the mathematical inequality $(r_1 + r_2) > (r_{12} + r_B)$ was presented. Why is the background count rate added to r_{12}?
5. Background count rates were not considered in Experiment 1. If they were, would the value obtained for the operating voltage change significantly?
6. A 30-minute count of 1.0 L of tap water gives 2,344 counts. A 30-minute background count gives 547 counts. What are the net count rate and the percent relative error for these data? These numbers are based on real measurements. The nuclide responsible for almost all of the radioactivity in our drinking water is naturally occurring ^{222}Rn.

7. When very low background count rates are needed for a particular
   measurement, steel cast before 1945 is often used. Why would it be
   important to use such old steel?

chapter four

Experiment 4 – inverse square law

The objectives of this experiment are to observe the relationship between the intensity of the radiation and the distance between the source and the detector and to determine if the inverse square law is applicable to the observations.

The inverse square law relates the intensity of radiation with the inverse of the square of the distance, $I = k/d^2$. The inverse square law is an important concept in personal protection. In addition, increasing the distance between the source and the detector lowers the count rate, thereby reducing the error associated with resolving time loss, and decreasing the distance between the source and the detector raises the count rate, thereby reducing the statistical error.

Procedure

a. Use a Geiger-Müller detector in a tube stand, as shown in Figure 10, or some other apparatus that allows convenient measurement of the distance from the radiation source to tube window.
b. Position a sealed "hard" beta source at a measured distance away from the detector, and measure the activity of the source. Record both the count rate and the distance from the source to the detector.
c. Move the source toward the detector. Measure the distance between the source and the detector, and measure the activity of the source at this distance. Record the distance and the activity.
d. Repeat step (c) several times at other distances.

Experimental data

Distance Count Rate	Distance Count Rate	Distance Count Rate

Figure 10 Experimental investigation of the inverse square law.

Report

a. Using Microsoft Excel™ or similar software, plot the observed count rate against the reciprocal of the square of the distance between the radioactive source and the Geiger-Müller detector.

b. Comment on any deviations from a linear relationship.

Questions

1. The dose rate 1.0 cm away from a radium seed is 2.35 rad/h. With what length forceps should this radium seed be handled to reduce the dose rate to the hand to 10 mrad/mrad?

2. Using a Geiger-Müller detector, a radiologist measured the dose rate of 240 mrad/h at a distance of 10.0 m from a ^{60}Co source. What

dose would she receive while performing a 20-minute procedure at a distance of 0.5 m from this source?

3. In addition to increasing distance, what other measures can be taken to lower doses when working with radioactive materials?

4. Why is a hard beta source used in this experiment? Could other radioactive sources be used? Briefly explain the answer.

5. What are the significant sources of error in this experiment? Are these errors the same for all data points on your graph? Explain.

chapter five

Experiment 5 – corrections for geometry factors

The objectives of this experiment are to examine the effect of sample position on the observed count rate, and to determine the efficiency of a Geiger-Müller counter for measurements of radioactivity.

Rarely, if ever, does one radioactive disintegration give rise to one event in a typical Geiger-Müller counter. The efficiency of a typical Geiger-Müller counter is between 1 and 10%. Some of the factors responsible for this are shown in Figure 11.

The distance between the sample and the detector is one such factor. The inverse square law predicts the intensity of radiation reaching the detector will decrease as the square of the distance between the detector and the source of radiation is increased. Radiation from a point source is emitted spherically in all directions. The counting geometry, G, is the ratio of the area of the surface of the sphere covered by the detector to the total area of the radiation sphere. The area of the surface of the sphere covered by the detector is essentially the area of the detector window, which is a constant. The total surface area of the radiation sphere is $4\pi r^2$. Hence, the geometry is inversely proportional to the square of the distance between the source and the detector, $G \propto 1/r^2$. Constant counting geometry is necessary in order to meaningfully compare the results obtained with different samples.

Procedure

Measure the activity of a beta emitter at each shelf under a Geiger-Müller counter, and repeat the measurements using other beta emitters.

Experimental data

		Activity	
Shelf	Source _____	Source _____	Source _____
1			
2			

3 _____

4 _____

5 _____

6 _____

7 _____

8 _____

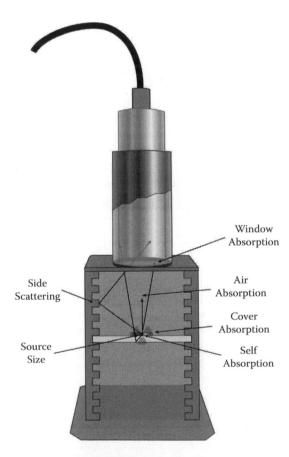

Figure 11 Factors influencing the observed count rate.

Report

a. Using the count rate observed on the shelf closest to the detector as the reference shelf, calculate the shelf ratios by dividing the count rate on a particular shelf by the count rate observed on the reference shelf.

b. Comment on the similarities and differences of the shelf ratios for the beta emitters.

Questions

1. Why is it relatively unlikely that a single radioactive decay in a source will give rise to one event inside a G-M detector?

2. Examine the factors illustrated in Figure 11. Which will increase count rates? Which will decrease count rates? Ignore sample size in answering this question.

3. How will increasing the sample size affect count rate? Assume a constant amount of radioactive material, but spread out more under the detector.

4. In addition to the inverse square law, what other factor may contribute to lowering of count rates as source-detector distance increases?

5. Qualitatively, what shelf ratios would you expect from an alpha-emitting nuclide? How about a purely gamma-emitting nuclide?

chapter six

Experiment 6 – back scatter of radiation

The objectives of this experiment are to examine the effect of back scatter on the observed count rate, and to relate the extent of back scatter with the atomic number of the medium responsible for the back scatter.

Radiation from a radioactive source can be scattered into or away from the detector, leading to increases or decreases in the observed count rates. Inspection of Figure 11 (Chapter 5) shows back scatter (and side scatter) deflects radiation into the detector, while fore scatter or air scatter deflects radiation away from the detector. Consideration of these factors is necessary to make meaningful measurements.

Boyarshinov (1967) reported experimental data showing the intensity of back-scattered radiation from ^{106}Ru was dependent on the atomic number of the scattering medium: $I = kZ^n$, where n, usually taken to be 0.67, varied from 0.45 to 0.80. This is shown in Figure 12. Soller et al. (1990) have applied back scatter measurements to the determination of atomic number using ^{90}Sr/^{90}Y beta radiation, and Johnson et al. (1992) have calculated the dose to the marrow due to back scatter at the bone-marrow interface for beta radiations from ^{153}Sm, ^{186}Re, and ^{166}Ho.

A more practical consideration of back scatter effects is avoidance of mixing planchettes—aluminum, glass, plastic, stainless steel, etc.—when preparing samples for comparative measurements. The average atomic numbers based on weighted average compositions for glass, plastic, and stainless steel are 12, 3, and 26, respectively. Relative to aluminum (or glass) planchettes, back scatter is more from stainless steel planchettes and less from plastic planchettes.

Procedure

Count a film-mounted source on several different backing materials.

Experimental data

Backing Material	Aluminum	Copper	Iron	Lead
Activity, cpm	_____	_____	_____	_____

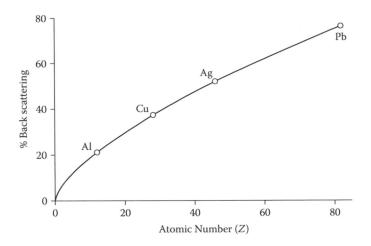

Figure 12 Back scatter of radiation.

Report

a. Using Microsoft Excel™ or similar, plot count rate against the atomic number of the backing material.
b. Comment on the relationship, if any, between scatter and atomic number of the backing material.

Questions

1. What is back scatter? Could it have had a significant effect on any of the previous experiments?
2. Two identical radioactive sources are counted on aluminum and steel planchettes. Will they give the same (within statistical error) count rates? Explain.
3. Would beta particles with different energies than those used in this experiment give different results? Briefly explain.
4. How does thickness of the backing material affect the results of this experiment? Is there a limit?
5. Would similar back scattering percentages be expected with alpha particles? How about gamma radiation?
6. What are the ideal detector conditions for a pure beta-emitting nuclide like ^{32}P?

chapter seven

Experiment 7 – corrections for self-absorption

The objective of this experiment is to examine the influence of sample thickness on the observed count rate.

The observed count rate depends upon many factors in addition to the number of radioactive atoms in the sample being measured. Self-absorption is only one such factor.

Consider a sample of $Ba^{14}CO_3$ having an activity of 500 Bq/mg. For a 1.0 mg sample spread uniformly over an aluminum planchette and counted under a Geiger-Müller detector with 1.0 % efficiency, the observed count rate will be 300 cpm. A 100-mg sample of the same material measured under the same conditions will have an observed activity less than the expected 30,000 cpm. Essentially, all of the radiation escapes from the thin 1.0 mg sample. When the thickness of the $Ba^{14}CO_3$ is increased, some of the radiations from the bottom of the sample are absorbed before escaping from the solid. This is an example of self-absorption.

As shown in Figure 13, self-absorption decreases the observed count rate as the sample thickness increases up to a point where only the radiation originating at and near the surface of the sample is detected. At this point the sample is said to be infinitely thick.

Self-absorption continues to be a problem in measuring the gross alpha activity of solid residues obtained from the evaporation of water samples (Dickstein et al., 2008). Many of these difficulties are resolved using liquid scintillation counting (Ruberu et al., 2008). Liquid scintillation counting has eliminated many of the empirically determined corrections associated with the measurement of ^{14}C activity in solid $BaCO_3$ (Gora and Hickey, 1954; Hendler, 1959; Wood, 1971).

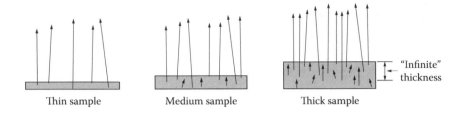

Figure 13 Self-absorption.

Part A

Procedure

   a. Pipette 1.0 ml of 10% (m/v) uranyl nitrate solution into a small petri dish, and measure the activity for 5 minutes.
   b. Add a second milliliter of the 10% (m/v) uranyl nitrate solution to the petri dish, and again measure the activity.
   c. Continue the additions and measurements five more times.

Data

Total Volume Activity	Total Volume Activity	Total Volume Activity
1.0 ml _____	4.0 ml _____	7.0 ml _____
2.0 ml _____	5.0 ml _____	
3.0 ml _____	6.0 ml _____	

Report

   a. Using Microsoft Excel™ or similar, plot count rate against total volume.
   b. Explain the relationship between count rate in cpm and total activity in mg $(UO_2)(NO_3)_2$.

Part B

Procedure

   a. Pipette 1.0 ml of solution containing 10% (m/v) uranyl nitrate solution into a small petri dish, and measure the activity for 5 minutes.
   b. Add 1.0 ml of water to the petri dish, and again measure the activity.
   c. Continue the additions of 1 ml increments of water and the measurements of activity five more times.

Data

Total Volume Activity	Total Volume Activity	Total Volume Activity
1.0 ml _____	4.0 ml _____	7.0 ml _____
2.0 ml _____	5.0 ml _____	
3.0 ml _____	6.0 ml _____	

Report

    a. Using Microsoft Excel™ or similar, plot count rate against total volume.

    b. Explain the relationship between count rate in cpm and total activity in mg $(UO_2)(NO_3)_2$.

Questions

1. Calculate the specific activity (Bq/mg) of pure $Ba^{14}CO_3$ assuming all carbon present is ^{14}C and that no other radioactive nuclides are present.
2. Calculate the specific activity of the uranyl nitrate solution used in this experiment. Assume all uranium present is ^{238}U, and that no other radioactive nuclides are present.
3. Using the answer to 2, estimate the expected count rate for the detector for 1.0 ml of uranyl nitrate. Is the measured value significantly different from the specific activity calculated in 2? If it is, briefly explain why.
4. Rationalize any differences observed between Parts A and B of this experiment.
5. Two student groups count equivalent mass samples of KCl. Group A leaves the KCl in an "anthill" pile in the middle of the planchette, while group B carefully spreads the salt out in the planchette. Will the groups observe different count rates? Explain.
6. Why is self-absorption a significant issue for counting alpha-emitting nuclides and ^{14}C?

chapter eight

Experiment 8 – range of beta radiations

The objectives of this experiment are to measure the attenuation of beta radiations in aluminum absorbers and to relate the range of the radiation in aluminum with its energy.

Neutrino emission accompanies the radiation emitted in beta decay, and the energy associated with the transition from less stable to more stable is partitioned between both. Positron emission, β^+, is accompanied by emission of the neutrino, ν. The antineutrino accompanies negatron emission, i.e., $^{32}P \rightarrow {}^{32}S + \beta^- + \bar{\nu}$. E_{max} is 1.71 MeV, but very little of the beta radiation from the decay of ^{32}P is of maximum energy. Most are closer to 0.55 MeV, and the average energy of the antineutrino is approximately 1.15 MeV. Consequently, beta radiation shows a spectrum of energies ranging from zero up to a characteristic maximum, E_{max}. The average energy of the beta radiation is approximately ¼ to ⅓ E_{max}.

The range of beta radiation is related to the energy of the beta radiation. Hence, the measurement of the maximum range allows approximation of E_{max}, which is a characteristic of the radionuclide. The relationship between maximum range and maximum energy shown in Figure 14 was reported by Katz and Penfold (1952).

The Feather analysis (Feather, 1938) was developed as a method for determining the range of beta radiation by comparing its attenuation with aluminum foils relative to the attenuation of the beta radiation from ^{210}Bi. Aluminum absorbers were selected to minimize bremsstrahlung production. RaDEF, a secular equilibrium mixture of ^{210}Pb, ^{210}Bi, and ^{210}Po, was selected as the reference source in which the ^{210}Bi appears to have the half-life of its ^{210}Pb precursor. Radiations from the ^{210}Pb and ^{210}Po are absorbed by the aluminum foil covering the RaDEF reference source. Feather adopted the accepted value of 476 mg/cm^2 as the maximum range of ^{210}Bi beta radiation in aluminum. Others (Chase and Rabinowitz, 1962) recommended a value of 505 mg/cm^2.

The Feather analysis began with measuring the attenuation of beta radiation through increasing thicknesses of aluminum. The attenuation data were presented in semilogarithmic form, and the ^{210}Bi beta radiation attenuation was extrapolated to reflect a range of 476 mg/cm^2. The range of the ^{210}Bi radiation was divided into tenths, and the fraction of

27

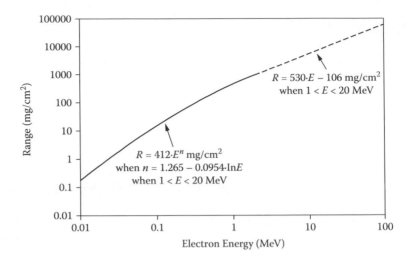

Figure 14 Beta range beta energy relationships.

the radiation transmitted at each tenth of the range was calculated. These transmission factors were then applied to the initial count rate of the beta emitter whose range was being determined, and its apparent range was read from its attenuation curve at each tenth. The apparent range was extrapolated to the final or tenth tenth.

The primary mode of interaction for beta radiation is with the orbital electrons of the medium through which it travels. The energy of the beta is dissipated in the medium it transverses in multiple collisions, each of which forms an ion pair. The number of ion pairs formed depends on both the energy of the beta radiation and the ionization energy of the medium.

To a lesser extent, the energy of beta radiation can be attenuated by interactions with atomic nuclei. Penetration of the orbital electrons brings the beta in the proximity of the intense electromagnetic field of the atomic nucleus. Here, energy is lost by the electron. The analogy would be to loss of kinetic energy and a consequent slowing of the beta. The energy loss in such interactions is referred to a bremsstrahlung production. The presence of brensstrahlung is shown at the right side of the attenuation curve in Figure 15. "Brensstrahlung" is a German word meaning braking radiation or radiation produced when the beta slows. The likelihood of bremsstrahlung production increases with beta energy and the atomic number of the absorber. Consequently, the Feather analysis is conducted using aluminum rather than lead absorbers.

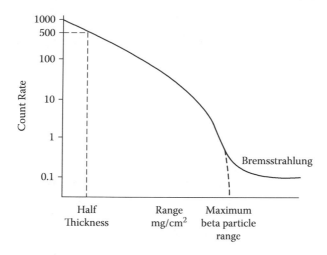

Figure 15 Beta attenuation.

Procedure

Measure the activity of a RaDEF reference source and several other beta emitters through increasing thicknesses of aluminum.

Experimental data

Absorber thickness	Activity of standard	Activity of unknown 1	Activity of unknown 2

Report

    a. Using Excel™ or similar software, plot the data as shown in Figure 15.

    b. Approximate the beta ranges using the method of Feather (1938) described in Appendix 3.

    c. From the relationship reported by Katz and Penfold (1952), estimate the beta energies.

    d. Compare the experimental values with those in the literature.

Tenth of Range	RaDEF Activity	Transmission Factor (TF)	Unknown 1 Unknown 1 × TF	Apparent Range	Unknown 2 Unknown 2 × TF	Apparent Range

Range of unknown 1 E_{max} of unknown 1

Beta emitter _____ mg/cm²   Beta emitter _____ MeV

Range of unknown 2 E_{max} of unknown 2

Beta emitter _____ mg/cm²   Beta emitter _____ MeV

Questions

    1. Why isn't lead used as an absorber in this experiment?

    2. Why are different range values obtained for different nuclides?

    3. Calculate the ranges in centimeters for the various beta-emitting nuclides studied.

    4. Select a nuclide from Appendix 2 that decays via negatron emission, taking care not to select one studied in this experiment. Estimate its maximum beta particle range in aluminum.

    5. How does neutrino emission complicate this experiment?

chapter nine

Experiment 9 – absorption of beta radiation

The objective of this experiment is to observe the absorption of beta radiation by different materials, to approximate the linear ranges of beta radiations emitted from several radionuclides in these materials, and to relate these ranges to both beta energy and atomic number of the absorber.

As stated above, the primary mode of interaction for beta radiation is with the orbital electrons of the medium through which it travels. The energy of the beta is dissipated in the medium it transverses in multiple collisions, each of which forms an ion pair. The number of ion pairs formed depends on both the energy of the beta radiation and the ionization energy of the medium. The number of collisions resulting in ion pair formation depends also on the electron density of the medium. Consequently, materials of high atomic number are more efficient absorbers than are materials of low atomic number.

Procedure

a. Measure the activity of a beta emitter through increasing thicknesses of thin aluminum, copper, and lead foils.

b. Repeat the measurements using beta emitters having higher and lower E_{max}.

Experimental data

	Beta emitter		
Absorber	_____	_____	_____
_____ mm Al	_____	_____	_____
_____ mm Al	_____	_____	_____
_____ mm Al	_____	_____	_____
_____ mm Cu	_____	_____	_____
_____ mm Cu	_____	_____	_____
_____ mm Cu	_____	_____	_____
_____ mm Pb	_____	_____	_____
_____ mm Pb	_____	_____	_____
_____ mm Pb	_____	_____	_____

Report

    a. Using Excel™ or similar software, plot the logarithm of the activity against absorber thickness for each nuclide.

    b. Calculate the linear absorption coefficient from the slopes for each of the plots in step (a).

    c. Using Excel™ or similar software, plot the absorption coefficients against the energy of the beta radiation.

    d. Using Excel™ or similar software, plot the absorption coefficients against the atomic number of the absorber.

    e. Comment on the relationships, if any, between absorption coefficient and energy, and absorption coefficient and atomic number.

Questions

    1. Why do the absorption coefficients vary when the absorber material changes?

    2. Why do the absorption coefficients vary when the energy of the beta particle changes?

    3 Two other ways to quantify the attenuation of radiation by matter are the mass (μ_m) and atomic (μ_a) attenuation coefficients. They can be calculated using the formulas below:

$$\mu_m = \frac{\mu}{\rho} \qquad \mu_a = \frac{\mu \times \text{atomic mass}}{\rho \times \text{Avogadro's \#}}$$

where μ is the absorption coefficient, ρ is the density of the absorbing material, and atomic mass refers to the absorber. Calculate the values of μ_m and μ_a for all three of the absorbers used in this experiment. Repeat for the other beta-emitting nuclides used. How do these attenuation coefficient values compare with each other? Explain any similarities or differences among these values.

chapter ten

Experiment 10 – absorption of gamma radiation

The objectives of this experiment are to measure the attenuation of gamma radiations by aluminum and by lead, and to calculate the linear and mass absorption coefficients and the half value layers for each of them.

The probability of interaction for gamma radiations is significantly less than that for alpha and beta radiations. Consequently, alpha and beta radiations are more easily stopped than gamma radiations. The range of the 7.69 MeV alpha radiation from ^{214}Po is 0.05 mm in aluminum, while approximately 3 mm of aluminum is required for the absorption of the 1.71 MeV (E_{max}) beta radiation from ^{32}P. More than 250 mm of aluminum would be required to absorb *half* of the gamma radiation from ^{137}Cs ($E_\gamma = 0.662$ MeV). The attenuation of gamma radiations is characterized by the half value layer, the thickness of absorbing material required to decrease the intensity of gamma radiations by a factor of 2 (see Bryan, 2009, p. 94; Loveland et al., 2006, p. 520). As shown in Figure 16, the half value layer is dependent upon both the gamma energy and the material in which it is absorbed.

Gamma radiations interact with matter in several ways. For photons with energies between 0.1 and 3 MeV, the energies usually associated with radioactive decay, interactions can be by the photoelectric effect, by the Compton effect, and by pair production.

The most likely mode of interaction for photons at the lower end of the range, 0.1 to 0.5 MeV, is by the photoelectric effect. The energy of the photon is absorbed completely by the target atom, and an electron, the photoelectron, is ejected from the target atom. The energy of the photoelectron is equal to the energy of the gamma photon less the electron binding energy. When the photoelectron is an inner shell electron, x-rays characteristic of the target atom frequently accompany the photoelectric effect.

Photons of energy between 0.5 and 1.0 MeV most likely interact by the Compton effect. In the Compton effect, only a portion of the gamma energy is absorbed by the target atom in an ionizing event. Some consequences of the Compton effect are described in Experiment 19.

In the intense electromagnetic field of the atomic nucleus, photons with energies in excess of 1.02 MeV can undergo an energy-to-mass transition and become a positron-negatron pair. The theoretical threshold for

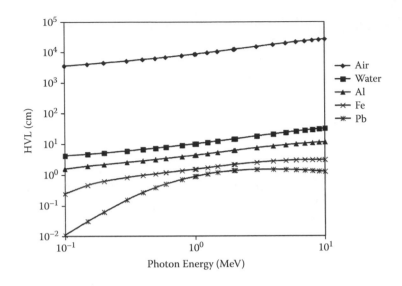

Figure 16 Half value layers for gamma radiation.

pair production is 1.02 MeV, the energy equivalent of two electron masses. Photon energy in excess of 1.02 MeV becomes the energy of the positron and the negatron, which undergo their typical interactions with matter culminating in annihilation radiation.

Procedure

Measure the activities of several gamma emitters through increasing thicknesses of aluminum and through increasing thicknesses of lead.

Experimental data

Aluminum absorbers		Lead absorbers	
Absorber thickness	Activity	Absorber thickness	Activity

Report

a. Using Microsoft Excel™ or similar software, plot the logarithms of the count rates against the linear thickness of the absorbers.
b. Calculate the linear absorption coefficients from the slopes of the lines.
c. Calculate the corresponding half value layers.
d. Comment on and explain any trend between the value for the half value layer and energy of the gamma radiation, and any trend between the value for the half value layer and the density of the absorber.

	Aluminum	Lead
Half value layer	_____ cm	_____ cm
Attenuation coefficient	_____ cm^{-1}	_____ cm^{-1}
Attenuation coefficient	_____ cm$^3 \cdot$ g^{-1}	_____ cm$^3 \cdot$ g^{-1}

Questions

1. A student observes 267 counts over 30 seconds for a high-energy gamma (>1 MeV) source by itself under a G-M detector. When the student places a thin piece of aluminum between the source and the detector, the count rate is observed to be 306 counts per 30 seconds. Assuming the experiment was performed correctly, explain how this result could occur.
2. Why is the probability of interaction lower for gamma rays than for alpha and beta particles?
3. Explain any similarities or differences in the attenuation coefficient values calculated in this experiment.
4. Using the formulas given in the previous experiment, calculate the mass and atomic attenuation coefficients for each of the gamma sources for both Al and Pb. Explain any similarities or differences observed in these values.
5. What are the most likely ways the gamma photons used in this experiment will interact with matter?
6. Assume this experiment was performed correctly using two sources of the same nuclide. The only difference is that one source has twice the activity of the other. Would the same or different values for the half value layer and the attenuation coefficient for these two sources be expected? Briefly explain.

chapter eleven

Experiment 11 – radioactive decay and instrument efficiency

The objectives of this experiment are to observe the decay of a radioactive nuclide, correct the data for instrument efficiency, and determine the half-life of the radioactive nuclide.

One characteristic of a radionuclide is that it undergoes radioactive decay. The rate of decay and the mode of decay for a given radionuclide are characteristic for that nuclide. The rate of an independently decaying radionuclide obeys first-order kinetics; i.e., the rate of decay depends only on the number of radioactive atoms:

$$-dN/dt = \lambda N, -\int dN/N = \lambda \int dt, -ln\ N = \lambda t + c$$

When $t = 0$, $N = N_0$, and when $t = t$, $N = N_t$, where N_0 is the number of radioactive nuclides present at time zero, N_t is the number of radioactive nuclides remaining after time t, λ is the decay constant for the radioactive nuclide, and t is the duration of time during which the decay took place. When $t = 0$, $c = -ln\ N_0$, and when $t = t$, $-ln\ N_t = \lambda t - ln\ N_0$. Then:

$$ln\ N_0 - ln\ N_t = \lambda t = ln\ (N_0/N_t),\ or\ -\lambda t = -ln\ (N_0/N_t),\ or\ N_t = N_0 e^{-\lambda t}$$

Half-life is the time during which half of the radionuclides undergo decay. After one half-life of radioactive decay, $t = t_{1/2}$, and $N_t = N_0/2$. Then $\lambda t_{1/2} = ln\ (N_0/N_0/2)$, and $\lambda t_{1/2} = ln\ 2$, or:

$$\lambda t_{1/2} = 0.693,\ \lambda = 0.693/t_{1/2},\ and\ t_{1/2} = 0.693/\lambda$$

Half-life has units of time, seconds, minutes, hours, days, years, etc. The decay constant, λ, has units of reciprocal time, second^{-1}, minute^{-1}, hour^{-1}, day^{-1}, year^{-1}, etc. It is the rate constant for the first-order decay reaction.

Periodic measurement of activity allows the determination of half-life when the radionuclide is neither very short lived nor very long lived. This approach works well for radionuclides having half-lives ranging from a few hours to a few weeks.

The background radiation from natural sources contributes to the results obtained in the detection and measurement of radioactive samples.

Consequently, it is customary to correct gross count rates by subtracting separately measured background count rates to obtain the net count rates of the samples. Background corrections are described and studied in Experiment 3.

Day-to-day variations in detection efficiency can be attributed to a variety of electrical and electronic sources. Compensations for these variations can be made by applying correction factors obtained from evaluations of the detection efficiency of a long-lived reference standard, such as ^{137}Cs or ^{90}Sr ($t_{1/2}$ = 30 and 28 years, respectively).

Procedure

At intervals recommended by the instructor, make measurements on (1) the background activity, (2) the activity of the half-life unknown, and (3) the activity of the long-lived standard.

Experimental data

Time/date	Background	Unknown	Standard

Report

   a. Construct a control chart from the net activity measurements made on the long-lived standard.
   b. Correct the half-life measurements as necessary.
   c. Using Microsoft Excel™ or similar software, plot the logarithm of the corrected count rates against time.
   d. Determine the half-life of the unknown from the slope of the curve: half-life: _____

Questions

1. What is the approximate age of the charcoal from a wood fire found in a cave showing an activity of 7.0 ± 1.3 dpm per gram of carbon?
2. A shipment of $Na_2^{51}CrO_4$ was certified to have a specific activity of 182 µCi/ml upon receipt at a clinical laboratory. What will be the specific activity of this material 2 weeks later?
3. Identify the product by its chemical symbol and mass number for each of the following radioactive decays:
    a.   The decay of ^{35}S by negatron emission
    b.   The decay of ^{222}Rn by alpha emission
4. Why couldn't the procedure used in this experiment be used for a nuclide with an especially short or especially long half-life?
5. ^{241}Am is commonly used in smoke detectors. One hundred and fifty-one years in the future an archeologist digs up a smoke detector that has an activity of 1.45×10^6 dpm. If the smoke detector originally had 0.870 µCi of ^{241}Am, in what year was the americium isolated to make this detector?
6. A rock is found to contain 0.492 g of ^{206}Pb and 0.789 g of ^{238}U. Assuming that all the ^{206}Pb comes from the decay of ^{238}U and that the half-lives of all other nuclides in the decay series are short compared to ^{238}U, how old is this rock?

chapter twelve

Experiment 12 – half-life determination

The half-life ($t_{1/2}$) of a radionuclide can be determined by observing the changes in decay rate over time. The time it takes for this value to be halved is the half-life. The best way to determine the half-life is to take advantage of the following mathematical relationships, where t is time, A is activity (or count rate), and λ is the decay constant:

$$t_{1/2} = \frac{\ln 2}{\lambda} \text{ and } \ln A = -\lambda t + \ln A_0$$

From the second equation it is apparent that a plot of $\ln A$ vs. t will yield a straight line with slope = $-\lambda$. Dividing the decay constant into $\ln 2$ will then give the half-life.

If there is sufficient activity and time, it is usually best to collect data over at least two half-lives. In this experiment, the half-life of ^{137m}Ba will be determined after it is separated from ^{137}Cs.

$$^{137}_{55}Cs \rightarrow {}^{137m}_{56}Ba + \beta^{-1} \qquad {}^{137m}_{56}Ba \rightarrow {}^{137}_{56}Ba + \gamma$$

Separation will be accomplished by passing a solution over an ion exchange column embedded with ^{137}Cs. The ^{137}Cs^{1+} ion sticks to this particular column, but once formed, the ^{137m}Ba^{2+} ion does not. As an eluent is passed over the column, the Ba^{2+} ions flow through, while the Cs^{1+} ions remain trapped. The decay of ^{137m}Ba can then be conveniently monitored without the presence of any ^{137}Cs. Radionuclide generator systems like this are commonly used in hospitals to produce the nuclides necessary for nuclear medicine applications.

Procedure

a. Place a 2.4-cm round filter paper into a planchet and carefully squeeze a drop or two of eluent from the ^{137m}Ba generator onto the filter paper.

b. Quickly transfer the planchet to a radiation counter. Count the sample for 30-second increments over about 8 minutes.

Experimental data

Time (min)	Total counts recorded	Gross counts for 30 s	Corrected counts (cpm)
0.5			
1			
1.5			
2			
2.5			
3			
3.5			
4			
4.5			
5			
5.5			
6			
6.5			
7			
7.5			
8			

Report

a. For each time interval, calculate the natural logarithm of the corrected count rate in cpm.
b. Plot $\ln A$ vs. time using Excel™ or other graphing software. Add a linear trend line and display its formula, making sure the slope is given to at least three significant figures.
c. Calculate the half-life and compare it to the literature value.

Questions

1. Why is the filter paper used in this experiment?
2. Why isn't the sample dried before counting?
3. Explain why the solution used in this experiment is a *minor* radiological hazard, i.e., why is the ^{137m}Ba solution relatively safe to work with?

4. Assuming the ^{137}Cs was also isolated from the solution used in this experiment, sketch a graph of ln cpm vs. time for parent, daughter, and total activity starting at the point of separation and ending 10 minutes later.

chapter thirteen

Experiment 13 – investigation of two independently decaying radionuclides

The objectives of this experiment are to observe the decay of a mixture of two independent radioactive nuclides, to resolve the composite decay curve, and to determine the half-lives of both radioactive nuclides.

In nature, silver exists as a mixture containing 51.8% ^{107}Ag and 48.2% ^{109}Ag. When natural silver is bombarded with neutrons, each of these stable nuclides can capture a neutron to form radionuclides. These two reactions may be summarized by the following:

$$^{107}_{47}\text{Ag} + ^{1}_{0}\text{n} \rightarrow ^{108}_{47}\text{Ag} + \gamma \qquad ^{107}\text{Ag (n,}\gamma\text{) } ^{108}\text{Ag}$$

$$^{109}_{47}\text{Ag} + ^{1}_{0}\text{n} \rightarrow ^{110}_{47}\text{Ag} + \gamma \qquad ^{109}\text{Ag (n,}\gamma\text{) } ^{110}\text{Ag}$$

The radioactive nuclides decay by beta emission to their stable isotopes of cadmium as follows:

$$^{108}_{47}\text{Ag} \rightarrow ^{108}_{48}\text{Cd} + \beta^- + \bar{\nu} \qquad t_{1/2} = 2.39 \text{ m}$$

$$^{110}_{47}\text{Ag} \rightarrow ^{110}_{48}\text{Cd} + \beta^- + \bar{\nu} \qquad t_{1/2} = 24.6 \text{ s}$$

The count rate (A) detected by the G-M detector is the sum of the count rate from each of the two nuclides. The decay of one nuclide does not influence the decay of the other. They are independently decaying activities. This can be expressed as their sum in the equation $A_{\text{total}} = A_{\text{Ag-108}} + A_{\text{Ag-110}}$, and shown in Figure 17.

The data will not fit a straight line, even when plotted on a semilog graph. The data points will fall along a curved line as shown in Figure 17 (square data points). This curve can be resolved into the two straight lines, representing the activities of each radioactive nuclide.

Because the half-life of ^{110}Ag is only 24.6 seconds, the sample must be transferred from the site of neutron bombardment to the site of activity measurement as quickly as possible. Set the counter to collect data every

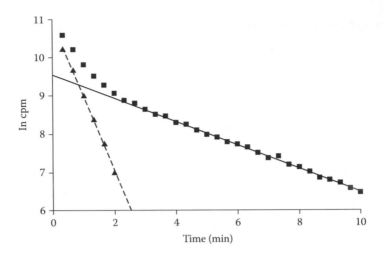

Figure 17 Activity of two independently decaying radioactive nuclides vs. time.

10 seconds. Collect data for 10 minutes and transfer them to an Excel™ spreadsheet. Correct the data for background, then coincidence as needed, paying special attention to the units.

As seen in Figure 17, the longer-lived ^{108}Ag is the major contributor to the total activity after approximately 200 seconds. If this experiment were modified so that the initial measurement of total activity was delayed by 3 minutes after irradiation of a piece of silver ended, no information about the shorter-lived ^{110}Ag would be obtained. The ^{110}Ag would have decayed through more than seven half-lives, making its count rate less than 0.8% of its initial count rate. For all practical purposes, the contribution to the total count rate from ^{110}Ag is negligible after 3 minutes. At this point in time, the longer-lived ^{108}Ag ($t_{1/2}$ = 2.39 m = 143 s) has decayed for less than 1.5 half-lives and still has 35% of the original ^{108}Ag activity. Therefore, the experimental data collected more than 3 minutes after the initial measurement can be treated as the data of a single-component decay. These data should be a straight line with a slope corresponding to the decay constant of the long-lived nuclide. The half-life can be calculated from the decay constant.

Procedure

   a. Radioactivate a silver foil (or a pre-1964 U.S. 10¢ piece) in the neutron source for 10 minutes.
   b. Remove the foil from the neutron source, and note the time of removal.

   c. Place the radioactivated silver foil under a Geiger-Müller detector, and begin the measurement of radioactivity at a time corresponding to 10 seconds after removal from the neutron source.

   d. At 10-second intervals, record the total number of counts accumulated on the scaler.

Experimental data

Time after removal	Total counts recorded	Gross counts for 10 s	Corrected counts for 10 s
0	0		
10			
20			
30			
40			
50			
60			
70			
80			
90			
100			
110			
120			
130			
140			
150			
160			
170			
180			
190			
200			
210			

220 _____

230 _____

240 _____

250 _____

260 _____

270 _____

280 _____

290 _____

300 _____

310 _____

320 _____

330 _____

340 _____

350 _____

360 _____

370 _____

380 _____

390 _____

400 _____

410 _____

420 _____

430 _____

440 _____

450 _____

460 _____

470 _____

480 _____

490 _____

500 _____

510 _____

520 _____

530 _____

540 _____

550 _____

560 _____

570 _____

580 _____

590 _____

600 _____

Report

a. First calculate the natural logarithm of the corrected cpm data.

b. Plot ln A vs. time t from about 3 to 10 minutes. Another range can be used if it is warranted. Determine a linear trend line and have Excel™ give the equation and R^2 of the line to three significant figures. The slope of this plot is equal to $-\lambda$, the decay constant. The half-life can be calculated from λ by using the equation

$$t_{1/2} = \frac{\ln 2}{\lambda}$$

At best, the count rates are valid to two or three significant figures so that the calculated half-life must be rounded off to the same number of significant figures.

c. Since the decays are independent of each other, the straight line formed by the ^{108}Ag data can be extrapolated back to time zero. If you use Excel, adjusting the Forecast values under the Options tab in the Format Trendline window can do this. This gives a graphical way to see the contribution to total count rate of the longer-lived ^{108}Ag from 0 to ~3 minutes. The distance from the extended trend line up to the total count rate data represents the count rate of the short-lived nuclide.

d. The count rate of the short-lived nuclide can also be calculated. Use the equation for the line to calculate the initial ^{108}Ag count rate for the data points during the first 3 minutes. Convert this number back to cpm by taking the natural antilog (e^x). Now calculate counts due to ^{110}Ag for each one of these points by subtracting the ^{108}Ag count rate (cpm) from the total count rate (cpm). Finally, take the natural log of the ^{110}Ag count rate. The formula function built in to Excel can be used to prepare a tabulation of the observed and calculated data. Any negative numbers are a result of uncertainties in the counting data and should be disregarded.

e. The data are now resolved into the two independently decaying nuclides. Plot the calculated data for the ln^{110}Ag count rate, and determine slope, R^2 value, and half-life. Since ^{110}Ag data are obtained by the subtraction of two activities, the best data are obtained when they are dramatically different from each other (0 to ~2 minutes).

f. Compare the results with the literature values and calculate the percent error. What factors could have affected the half-lives? Be specific!

Questions

1. What percentage of the original activity is left after a radionuclide decays through one half-life? After seven half-lives? After 10 half-lives?
2. If Nd is placed in the neutron source, what radioactive isotopes might be formed? Sketch a plot of total count rate vs. time, labeling areas of the plot where the specific isotopes would be expected to contribute to the observed count rate.
3. As a practical matter, would it be possible to perform this experiment with gallium instead of silver? Why or why not?
4. To determine the ^{110}Ag count rate, the ^{108}Ag count rate in cpm was subtracted from the total count rate in cpm. Why not save a couple steps and subtract the ^{108}Ag count rate in cpm from the total count rate in cpm?
5. Using the data, determine the ^{110}Ag activity at 3 minutes. Is it negligible when compared to the total activity?
6. If an unknown element is irradiated with neutrons, how could it be determined whether one or more than one radioactive nuclide is produced?

chapter fourteen

Experiment 14 – half-life of a long-lived radionuclide

The half-life of ^{40}K is 1.277×10^9 years. There is no experimental method of detecting any change in the count rate of a potassium sample during our lifetimes. This means that the method used in determining the half-lives of ^{108}Ag and ^{110}Ag in Experiment 13 cannot be used here. However, we can use the following equation, as long as we know activity (A) and number of ^{40}K atoms (N):

$$A = \lambda N = \frac{\ln 2}{t_{1/2}} N$$

To calculate half-life, both activity and number of atoms must be determined. Activity can be determined by counting a sample of KCl (cpm) then calculating the decay rate (dpm) from the experimentally determined counter efficiency. Calculation of activity is complicated by the branched decay of ^{40}K as illustrated in Figure 18. ^{40}K is located between the stable isobars of ^{40}Ar and ^{40}Ca, suggesting it can decay in either direction to achieve stability. There is sufficient energy for positron decay ($E = 1.505$ MeV), but the main pathway to ^{40}Ar is via electron capture to an excited state of ^{40}Ar followed by the emission of a 1.461 MeV γ ray.

The branch ratios are given as percentages in Figure 18. Notice that beta particles and gamma rays are parts of two different decay modes. In the decay of 100 atoms of ^{40}K, there will be approximately 89 beta particles and 11 gamma rays emitted. The gamma rays will not have a significant impact on the count rate if a G-M detector is used for this experiment. A calculated half-life from beta counting will be higher than the correct value. The half-life calculated using data from only one branch of the decay is called a partial half-life. In *all* cases the partial half-life is longer than the true or total half-life. If the half-life is calculated using beta counting as 1.44×10^9 years, then multiplying by 0.8928 (the branch ratio) will give the correct value.

The number of atoms of ^{40}K in the sample can be calculated from the mass of the sample and Avogadro's relationship of mass to atoms. Naturally

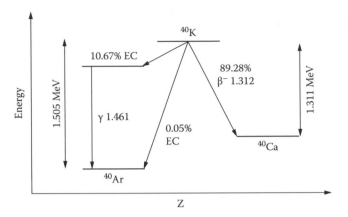

Figure 18 Decay scheme for ⁴⁰K.

occurring potassium has an average atomic mass of 39.0983 g/mole, and is a mixture of three isotopes. ³⁹K and ⁴¹K are stable, and ⁴⁰K is radioactive. ⁴⁰K is naturally occurring because of its very long half-life. The isotopic abundances are as follows: ³⁹K (93.2581%), ⁴⁰K (0.0117%), ⁴¹K (6.7302%).

Procedure

   a. Measure background counts for at least 30 minutes. Calculate the background count rate (cpm).
   b. Weigh out approximately 300 mg of a potassium salt (KCl works well) into a planchette.
   c. Being careful not to spill any, carefully spread the KCl out in the planchette.
   d. Count the sample under a G-M detector for at least 1 hour. Calculate the gross and net count rates.

Experimental data

Sample mass: _____

Gross counts: _____   Gross count rate: _____

Background counts: _____   Background count rate: _____

                                                   Net count rate: _____

Report

a. Calculate the number of ^{40}K ions present in the sample.
b. Calculate activity (decays per year) of the sample using counter efficiency.
c. Calculate the partial and true half-lives.
d. Compare the experimental value to the literature.

$N =$ _____

$A =$ _____

Partial $t_{1/2} =$ _____

True $t_{1/2} =$ _____

Questions

1. The average human carries around 16 kg of carbon, and has an activity of 2.5×10^5 dpm due to ^{14}C. Use these numbers to calculate the half-life of ^{14}C (% abundance $= 1.2 \times 10^{-10}$%).

2. A 30-minute count of 1.0 L of tap water gives 2,344 counts. A 30-minute background count gives 547 counts. If the detector efficiency is known to be 12.1% and no coincidence correction is required at this count rate, what is the activity (pCi/L) of this sample? These numbers are based on real measurements. The radionuclide responsible for almost all of the radioactivity in drinking water is ^{222}Rn.

3. Radiological workers are often subjected to testing to ensure they are not ingesting or absorbing radionuclides. One test is the whole body count. The worker is brought into a heavily shielded room and asked to lie still while detectors monitor the subject for radioactivity over 30 to 60 minutes. Even if the worker has not inadvertently ingested some of his or her work, some radiation is always detected. This could be due to naturally occurring ^3H, ^{14}C, or ^{40}K. Which, if any, of these nuclides are detected during a whole body count? Explain your answer.

4. Much of the shielding for rooms used for whole body counts is steel. It is important that steel used in these rooms be cast before 1945. Why would this make a difference?

5. Activity values determined in this experiment tend to be systematically low. Can you suggest a simple reason for this? How does this systematic error affect the half-life?

6. Why is it necessary to count for such long times in this experiment?

7. The average banana contains 450 mg of potassium. What is the activity of the average banana?

8. Why don't the gamma rays emitted in the decay of ^{40}K have a significant effect on the results of this experiment?

chapter fifteen

Experiment 15 – autoradiography

The objectives of this experiment are to master the techniques of autoradiography and to apply them to observing the translocation of a nutrient in a plant.

Nuclear radiations affect photographic emulsion in much the same way as does visible light. The energy of the radiation initiates reduction of the silver halide in the emulsion to produce a latent image. The greater the quantity of energy absorbed by the emulsion, the greater the extent of silver halide reduction. The smaller the particles of silver halide, the sharper the image. The larger the particles of silver halide, the faster their reduction. These factors influence the resolution and speed of photographic film. When the film is developed, the "hypo" (sodium thiosulfate) forms a soluble complex with the silver halide. This allows it to be washed out of the emulsion. The metallic silver formed in the reduction of the silver halide remains in the emulsion as many tiny black particles. The emulsion is "hardened" with dilute acetic acid, and the latent image is visible as a negative. Those regions of the emulsion absorbing the greatest quantities of radiation contain the greatest quantities of metallic silver particles and appear as the darkest regions of the negative. A typical autoradiogram is shown in Figure 19.

To prepare an autoradiogram, the specimen containing radioactive material is placed in close contact with the photographic emulsion. Optimal contrast is obtained with the absorption of 10^6 to 10^7 β/cm^2. The exposure time should be adjusted accordingly. This can be approximated from the activity measured with a Geiger-Müller detector 1 cm above the specimen covered with a lead sheet having a 1 cm² opening. Following exposure, the emulsion is processed in accord with the manufacturer's instructions under dark room conditions. The localization of the radioactive material in the specimen is determined from the pattern of darkening in the emulsion. Quantification is difficult. For gross specimens, the relative darkening can be measured densitometricly and compared to standard exposures.

The quality of the image depends upon many factors. Among them are grain size of the silver halide, thickness of the emulsion, thickness of the specimen, contact between the specimen and the emulsion (geometry), radiation type, radiation energy, and exposure time. Close contact between thin samples and thin emulsions of fine grain silver halide usually gives

Figure 19 Autoradiogram of ^{32}P distribution in a leaf from *Arabidopsis thaliana*. (From W. Paul Quick and Randy Taylor. www.biospacelab.com/htm/M1A_html. With permission.)

the best results. The ranges of some beta emitters in a typical photographic emulsion are ^{3}H ~ 2 μm, ^{14}C or ^{35}S ~ 100 μm, and ^{32}P ~ 3,000 μm.

Some typical applications of autoradiography have been in determining the absorption and utilization of foliar applied nutrients labeled with ^{32}P (Wittwer and Lundell, 1951), analyzing the transmembrane distribution of ^{125}I-labeled concanavalin A (Fisher, 1982), and measuring the binding of some ^{3}H-labeled cortical glutamatergic markers in schizophrenics and controls (Scarr, et al., 2005).

Procedure

a. Select small, broad-leaved, cultivated or wild plants for this experiment.
b. Carefully remove the plants from the soil without damaging their rootlets.
c. Gently wash the plants free of soil and surface contamination.
d. Secure test tubes to ring stands.

e. Add 5 ml of 1×10^{-3} M phosphate buffer (equal molar quantities of NaH_2PO_4 and Na_2HPO_4) solution containing 0.5 µCi/ml (= ~2 × 10⁴ Bq/ml) to each test tube.

f. Carefully place one plant in each test tube so that the roots are immersed in the carrier-tracer phosphate solution.

g. Allow the plants to remain in contact with the carrier-tracer phosphate solution for several hours or overnight.

h. Taking all precautions to avoid direct skin contact with the radioactive material and to prevent contamination of the work area, remove the plants from the test tubes, rinse the roots by repetitive dipping in a beaker of water, blot dry with a Kimwipe™, and place each plant on a separate sheet of Saran™ wrap.

i. Using micropipettes, remove 100 µl aliquots of the carrier-tracer phosphate solutions from each of the test tubes, transfer them to separate planchettes, dry them, and measure and record the radioactivity of each.

j. Saran wrap side down, place each plant on a separate sealed sheet of autoradiography film, and cover with another sheet of Saran wrap.

k. Secure the carrier-tracer phosphate solutions, and rinse water, glassware, and Kimwipes for radioactive waste management procedures.

l. Measure the radioactivity of the specimen covered with a lead sheet having a 1 cm² opening using a Geiger-Müller detector at a distance of 1 cm, and calculate the exposure time as described above.

m. Sandwich the films and the Saran-wrapped specimens between sheets of light cardboard, and compress them with bricks.

n. At the end of the exposure period, treat the Saran-wrapped specimens as radioactive waste.

o. Process the films in accord with the manufacturer's instructions under dark room conditions.

Report

a. Identify the plants used in this experiment by their Latin names.

b. Show step-by-step calculation of the exposure time from the measurement of radioactivity above the 1 cm² opening in the lead sheet covering the specimen.

c. Calculate the fraction of the nutrient transported to each square centimeter of leaf surface.

d. Describe and interpret the autoradiogram.

Questions

1. Why are there differences in the ranges of beta particles in photo-graphic emulsions?
2. Briefly describe an experiment, involving radioactive tracers, that would demonstrate that carbon atoms in CO_2 are incorporated into glucose via photosynthesis.
3. Why is it important for the plant to be as close to the film emulsion as possible?
4. How would the autoradiogram of the plant differ from an x-ray of the same plant?

chapter sixteen

Experiment 16 – calibration and operation of the electroscope

The objectives of this experiment are to observe the discharge of the electroscope by alpha, beta, and gamma radiations, and to calibrate the electroscope for specific alpha-, beta-, and gamma-emitting nuclides.

The electroscope, like the Geiger-Müller counter, consists of a cylindrical cathode containing a gas of low ionization potential, usually air, and a central anode. The potential difference between the cathode and the anode in the former is significantly less than it is in the latter. Hence, only the electrons from the primary ionizations are collected because there is no multiplication due to secondary and subsequent ionization events. A typical electroscope is shown in Figure 20.

The energy required to produce an ion pair in air is approximately 32 eV. Hence, an alpha emitted in the decay of ^{210}Po can produce as many as 165,500 ion pairs in air. A beta emitted in the decay of ^{32}P will form far fewer ion pairs for several reasons. The average energy is less: ⅓ of E_{max} = 1.71/3 MeV = 0.57 MeV, and 570,000/32 = 17,800 ion pairs. In addition, it is possible for the beta to escape the detector without initiating an ionizing event. The 0.36 MeV gamma emitted in the decay of ^{133}Ba will form even fewer ion pairs. Not only is the energy less, but the probability of interaction is much less.

Procedure

    a. Charge the electroscope, adjust the zero, and measure the discharge rate.

    b. Recharge the electroscope, again adjust the zero, and measure the discharge rate for an alpha emitter.

    c. Repeat the above using beta and gamma emitters.

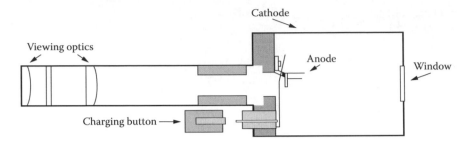

Figure 20 A typical electroscope.

Experimental data

		Time for a 10-Division Change		
Divisions	None	α Emitter	β Emitter	γ Emitter
00–10	_____	_____	_____	_____
10–20	_____	_____	_____	_____
20–30	_____	_____	_____	_____
30–40	_____	_____	_____	_____
40–50	_____	_____	_____	_____
50–60	_____	_____	_____	_____
60–70	_____	_____	_____	_____
70–80	_____	_____	_____	_____
80–90	_____	_____	_____	_____
90–100	_____	_____	_____	_____
Source used	_____	_____	_____	_____

Report

a. Calculate the discharge rates in terms of divisions per minute per μCi for the α, β, and γ emitters correcting for the background discharge as measured above.

α emitter _____   div/min/μCi of _____

β emitter _____   div/min/μCi of _____

γ emitter _____   div/min/μCi of _____

b. Compare and explain any differences in the discharge rates in terms of divisions per minute per μCi for the α, β, and γ emitters.

Questions

1. Calculate the number of primary ion pairs formed by one of the α particles emitted in the decay of ^{226}Ra to ^{222}Rn.
2. Why is ^{210}Po used to reduce static charge in the production of photographic film and phonograph records?
3. Oxygen has a tendency to react with electrons. Coupled with the fact that the electroscope does not have a large potential difference between cathode and anode, comment on the accuracy of this detector type.
4. Similar detectors are used in radiation therapy machines that produce high-intensity, high-energy x-ray beams. They are built into the machine directly in the beam. Why is this a good application of this detector type?

chapter seventeen

Experiment 17 – properties of proportional counters

The objectives of this experiment are to observe the dependence of the proportional counter response to alpha and beta radiations as a function of the potential applied to the anode, and to determine the operating potentials of the proportional counter for the measurement of alpha and beta radiations.

The proportional counter, like the electroscope and the Geiger-Müller counter, consists of a central anode surrounded by a gas of low ionization potential, often a 90% argon–10% methane mixture known as P-10. Unlike the electroscope, where there is little if any gas multiplication, and unlike the Geiger-Müller detector, where the gas multiplication factor is on the order of 10^{10}, 10^2 to 10^5 electrons are collected at the anode for each primary ion pair in the proportional counter.

Figure 21 shows a proportionality between the number of electrons collected and the number of primary ion pairs formed. As seen in Experiment 16, the specific ionization of alpha radiation is greater than that of beta radiation. Alpha radiation produces more primary ion pairs than beta radiation. With gas amplification, more electrons are collected from alpha radiation than from beta radiation. Consequently, alpha radiation produces larger electrical pulses than does beta radiation. It is possible to differentiate between pulses from alpha radiation and those from beta radiation. This makes it possible to measure alpha radiation in the presence of beta radiation.

As the potential at the anode is increased, the primary and secondary ionizations caused by the beta radiation undergo gas amplification to a point where they too produce electrical pulses of sufficient magnitude to produce counts. At such voltages, the pulses from both the alpha and the beta radiation contribute to the observed count rate.

The specific ionization of gamma radiation is small, and few primary ion pairs are produced. Gamma radiation makes little if any contribution to the observed count rate.

Figure 22 shows the proportional counter is capable of measuring alpha radiation in the presence of beta radiation from a sample of RaDEF. At applied voltages below approximately 1,400 volts, the alpha radiations from ^{210}Po make the major contribution to the observed count rate. At

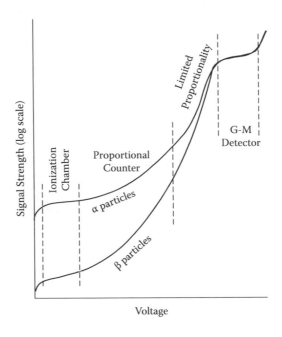

Figure 21 Gas multiplication of primary ionization.

applied voltages between 1,900 and 2,200 volts, gas multiplication of the primary ionizations from the beta radiation from the ^{210}Bi and ^{210}Pb is amplified sufficiently to produce detectable pulses, and both the alpha radiation and the beta radiation contribute to the observed count rate.

The dead time of a proportional counter is a few microseconds, more than 10 times less than that of a Geiger-Müller counter. Demands on high voltage stability are greater for the former. The anode in the former is frequently a thin tungsten wire loop.

In the 2π gas flow proportional counter with window shown in Figure 23a, the gas is replenished continually to ensure a constant supply of argon atoms for the primary and secondary and subsequent ionization events. The geometry of the detector makes approximately 50% of the radiation from the sample available for detection. Some loss is due to absorption by the thin detector window, and some gain (~2%) is due to back scatter from the stainless steel planchettes on which the samples are usually mounted. The terms 2π and 4π refer to the number of steradians around a point source in space. There are 2π steradians in a hemisphere and 4π in a sphere.

With the 4π windowless gas flow proportional counter shown in Figure 23b, the sample to be measured is deposited on an extremely thin membrane, which is then positioned between the upper and lower halves of

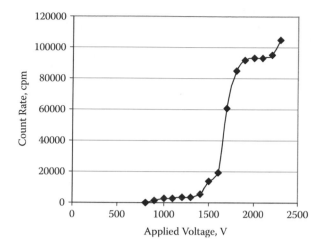

Figure 22 Characteristic voltage response of a proportional counter for RaDEF.

the chamber. Air is purged from the chamber by the P-10 gas prior to making the measurement. This arrangement approaches 100% counting efficiency.

In addition to finding applications to the measurement of alpha radiations, the sealed proportional counter has been applied to the measurement of x-rays.

Procedure

a. Place a RaDEF source in or under a proportional counter.
b. With the high voltage adjusted to 500 volts, make a 1-minute count, and record the result.

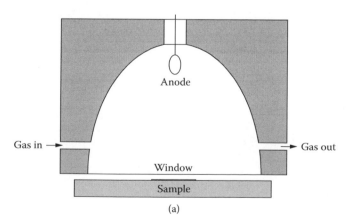

Figure 23a 2π proportional counter with window.

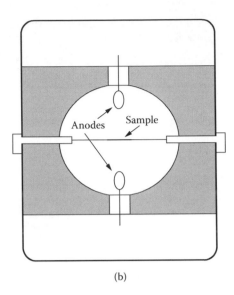

(b)

Figure 23b 4π windowless proportional counter.

    c. Increase the high voltage by 100 volts, and make another 1-minute
       count. Record this result.
    d. Repeat step (c) several times.

Experimental data

Voltage Count Rate	Voltage Count Rate	Voltage Count Rate

Report

    a. Using Microsoft Excel™ or similar software, plot the count rate, R,
       against the voltage, V.
    b. Identify the alpha and beta plateaus.
    c. Determine the operating voltages, V_O, for measuring alpha activity
       and total alpha and beta activity.
    d. Determine the alpha activity in cpm at the alpha operating voltage.
    e. Determine the beta activity in cpm at the beta operating voltage.

Questions

1. Calculate the number of electrons collected at the anode from a single primary ionization event in a proportional counter, and compare this number to the number of electrons collected at the anode from a single primary ionization event in a Geiger-Müller counter.
2. Can a G-M detector differentiate between alpha and beta radiations? Briefly explain the answer. Why aren't voltages below 500 V studied as part of this experiment?
3. If the contribution to the overall count rate by gamma photons emitted as part of alpha or beta decay is insignificant, how can proportional counters be used for x-ray measurements?

chapter eighteen

Experiment 18 – integral spectra

The objectives of this experiment are to observe the effect of applied voltage on the response of the sodium iodide detector to gamma radiations.

The data for this experiment are collected in a manner similar to that used previously in Experiment 1. The sodium iodide detector, however, is very different from the Geiger-Müller detector. The radiation, usually gamma, that passes through the aluminum covering and is absorbed in the sodium iodide (activated with 0.1% thallium iodide) crystal causes the excitation of electrons. Some of these electrons return to their original orbitals with the emission of photons (approximately 20 for every keV of gamma energy absorbed in the crystal). These photons are transmitted through an optical coupling to the photocathode of a photomultiplier tube (PMT). The absorption of these photons by the photocathode causes the ejection of photoelectrons (approximately 1 for every 10 photons absorbed), which are attracted to the first dynode of the PMT, where each causes the ejection of 4 or 5 electrons. These electrons are attracted to the second dynode state, where each causes the ejection of four or five electrons. The progression of dynode to dynode amplifies the electrical signal by a factor of 4 or 5×10^n, where n is the number of dynode stages. The magnitude of the electrical signal is proportional to the gamma energy absorbed in the sodium iodide crystal. The signal from a Geiger-Müller detector is independent of the beta energy absorbed.

The photocathode in the PMT has a dark current arising from the spontaneous and random emission of electrons. The PMT cannot distinguish between the photoelectrons due to scintillations and the spontaneously emitted thermal electrons. Both undergo multiplication. Consequently, the sodium iodide detector has high background due to these thermionic emissions in addition to the background from natural sources.

A sodium iodide scintillation crystal coupled to a photomultiplier tube is shown schematically in Figure 24.

Procedure

Measure the background activity and the activities of several gamma emitters at several high-voltage settings with the sodium iodide detector. Do not exceed 1,200 volts.

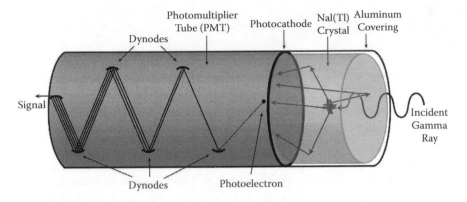

Figure 24 Sodium iodide–photomultiplier tube detector.

Experimental data

Gamma Emitter

Applied Voltage	_____ Activity	_____ Activity	_____ Activity	Background Activity
_____	_____	_____	_____	_____
_____	_____	_____	_____	_____
_____	_____	_____	_____	_____
_____	_____	_____	_____	_____
_____	_____	_____	_____	_____
_____	_____	_____	_____	_____
_____	_____	_____	_____	_____
_____	_____	_____	_____	_____
_____	_____	_____	_____	_____
_____	_____	_____	_____	_____
_____	_____	_____	_____	_____
_____	_____	_____	_____	_____

Report

a. Correct the data for background, and normalize to 1,000 cpm at 1,000 volts; i.e., determine a factor for each gamma emitter such that the activity at 1,000 volts will be 1,000 cpm when multiplied by this factor, and then multiply all of the corresponding measurements by the factor for that gamma emitter.

b. Using Microsoft Excel™ or similar software, plot the normalized activities for each gamma emitter against the applied voltage.

c. Compare the integral spectra on the basis of gamma energy.

Questions

1. Assume 20 luminescent photons are formed for every keV of gamma energy absorbed in a sodium iodide detector, and 95% of these are transmitted to the photocathode of a photomultiplier tube. Assume also that 1 photo-electron is ejected from the photocathode for every 10 luminescent photons absorbed at the photocathode. Finally, assume an amplification of 4×10^5 in the photomultiplier tube. From these assumptions, calculate the number of electrons collected at the anode of the photomultiplier tube when a gamma photon from ^{137}Cs is absorbed in the sodium iodide detector.

2. Repeat the calculation from question 1 for ^{133}Ba.

3. How do the background count rates observed in this experiment compare with those determined with a G-M detector? Briefly explain any difference.

chapter nineteen

Experiment 19 – gamma spectrometry I

The objectives of this experiment are to record and interpret the spectra of gamma-emitting radionuclides.

A review of the Compton effect and of gamma spectrometry may be helpful (Bryan, 2009). Were it not for losses of some Compton photons from the sodium iodide detector and randomness in photon emission and transmission in the sodium iodide detector, as well as randomness in the photoelectron ejection at the cathode in the photomultiplier tube (PMT) and in electron production at the dynodes, a gamma spectrum would consist of a single sharp line corresponding to E_γ. In Figure 25a, the gamma spectrum of ^{60}Co shows a continuum from approximately 500 to 1,000 keV and two peaks at approximately 1,150 and 1,300 keV, respectively. This Compton continuum results from the deposition of only a part of the gamma energy in the sodium iodide detector. The energy of the Compton electron was absorbed, but the Compton photon escaped. The two photopeaks result from total absorption of the 1.17 and 1.33 MeV gamma photons associated with the decay of ^{60}Co. The decay scheme of ^{60}Co is shown in Figure 25b. That the former photopeak is of greater inten sity than the latter reflects the energy dependence of the absorption coefficient of sodium iodide.

The peak-to-Compton ratio is a figure of merit and demonstrates the advantage of physically larger (and, consequently, more expensive) sodium iodide detectors.

Spectral resolution using the intrinsic germanium or the older lithium-drifted germanium (Ge(Li)) detector is far superior to that obtained with the sodium iodide, NaI(Tl), detector, as shown in Figure 26. The sodium iodide detector, however, has the advantage of greater efficiency.

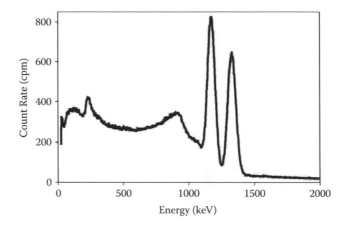

Figure 25a Gamma spectrum of ^{60}Co.

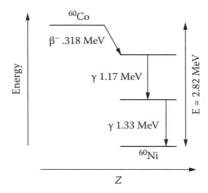

Figure 25b Decay scheme for ^{60}Co.

Procedure

a. Consult the manual or the instructor for instructions on the operation of the gamma spectrometer.
b. Calibrate the spectrometer so that a standard (^{137}Cs) gamma energy is centered in a channel number (33) that is an integer multiple of its gamma energy (662 keV).
c. Measure the spectra of other γ emitters.

Experimental data

Download or print the gamma spectra.

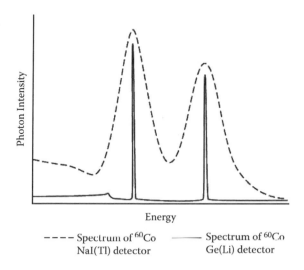

Figure 26 Spectrum of ^{60}Co; NaI(Tl) Detector; Ge(Li) Detector.

Report

a. Identify the peaks in the spectra.
b. Determine the energies and compare them to the literature values.
c. Determine the efficiency of the detector in terms of photopeak counts per μCi for each of the gamma emitters, and comment on any differences observed.

Questions

1. Could Experiment 13 be performed with the NaI detector instead of a G-M tube? Briefly explain.
2. How does the efficiency of the NaI detector compare with that of the G-M tube?
3. In theory, the amount of energy that can be deposited in the NaI detector via Compton scattering and subsequent escape of the scattered photon has an upper limit that is lower in energy than E_γ. Is there a sharp drop in the Compton continuum before the peak in your spectra? If not, explain?

chapter twenty

Experiment 20 – gamma spectrometry II

This experiment will facilitate identification of some of the features commonly observed in gamma spectra.

In general, gamma photons can interact with the matter inside and outside of the detector via Compton scattering or the photoelectric effect. These interactions can give rise to a variety of features that can be observed in gamma spectra. For an overview of the interactions and the types of spectral features created, see Bryan (2009).

As explained in Experiment 19, if a photon undergoes Compton scattering inside the detector and the scattered photon escapes the detector, then only part of its original energy is recorded. This gives rise to the Compton continuum, a broad, relatively flat feature that begins at an energy lower than that observed for the photopeak. The maximum energy that can be deposited in the detector crystal by Compton scattering can be found from the following equation:

$$E_{CE} = \frac{E_0^2}{E_0 + 0.2555}$$

where E_{CE} is the energy of the Compton edge and E_0 is the energy of the incident photon. Both carry units of MeV.

When a gamma ray passes through matter on its way to the detector, it can undergo Compton scattering. This will also shift counts from the photopeak to the Compton continuum and is termed object scattering.

A gamma photon originally headed away from the detector can undergo back scattering when it interacts with some of the matter in the room. While this does not have a high probability of occurring, it is common enough that a small back scatter peak is often observed in a gamma spectrum. To correctly identify a back scatter peak, calculate its energy using the following:

$$E_{BS} = \frac{E_0}{1 + (3.91 \times E_0)}$$

where E_{BS} is the energy of the back scatter peak in MeV. This peak often appears as a small bump on top of the Compton continuum.

The percent resolution is a measure of detector quality. If two gamma rays are too close in energy, they cannot be detected as separated peaks. The gamma spectrum of ^{57}Co has peaks at 122 and 136 keV, which are not separated from each other (not resolved). The gamma spectrum of ^{60}Co has peaks at 1.17 and 1.33 MeV, which should be resolved with a NaI detector. The percent resolution as calculated for the full width at half maximum (FWHM) of the peak is the usual method of determining this quantity. The equation for calculating the percent resolution is

$$\% \text{ resolution} = \frac{\Delta E_0}{E_0} \times 100\%$$

where ΔE_0 is the width of the photopeak halfway to its maximum intensity.

Procedure

   a. Record the spectra of several gamma-emitting radionuclides. Label any features observed.
   b. Calculate the value for the Compton edge for each photopeak on each spectrum. Mark those energy values on the spectra.
   c. Collect 1-minute spectra for a ^{137}Cs source placed right up against the detector, a plastic disc between the source and the detector, and a 1-cm Al block between the source and detector. Calculate the peak-to-total ratio for all three spectra. Comment on the observations.
   d. Calculate the energy of the back scatter peak expected for each photopeak observed. If they appear on the spectra, label them as such.
   e. Rerecord the spectrum of one of the nuclides that exhibits a back scatter peak. This time, place a piece of lead behind the source. Note any differences with the original spectrum.
   f. Using a spectrum that exhibits a clear photopeak, calculate the resolution. Do this by locating the channel number (energy) that has the highest count rate for the peak. Divide this count rate by 2. Locate this numerical value (as close as possible) on the downward slopes on either side of the peak. Note the channel numbers (or energy values) of these two points. Calculate the energy values for these two points. The FWHM (ΔE_γ) is the difference between these two values. Divide by the energy of the peak (E_γ) to obtain the resolution.

Questions

1. Calculate the energy (keV) of the back scatter peak and Compton edge for ^{54}Mn.
2. What would be the energy of an iodine escape peak in the gamma spectrum of ^{113m}Sn?
3. The resolution of the photopeak of the ^{137}Cs spectrum is 0.5% using a Ge(Li) detector. What is the FWHM? Why does it differ from what is observed with a NaI detector?
4. What is expected in the ^{18}F gamma ray spectrum? Briefly explain.

chapter twenty-one

Experiment 21 – liquid scintillation counting

The objectives of this experiment are to become familiar with the techniques of liquid scintillation counting, to determine the efficiency of liquid scintillation counting for ^{3}H and ^{14}C, and to observe and correct for quenching.

Low-energy beta radiations such as those from ^{3}H (E_{max} = 18 keV) and ^{14}C (E_{max} = 156 keV) are not efficiently measured with the sodium iodide detector or with the Geiger-Müller detector. The beta radiation from ^{3}H penetrates neither the aluminum cover surrounding the sodium iodide detector nor the window of the Geiger-Müller detector. The low-energy beta radiations from ^{14}C do not penetrate the aluminum cover surrounding the sodium iodide detector. Very few of them have sufficient energy to penetrate the window of the Geiger-Müller detector.

Beta radiations can be measured using scintillation detectors fabricated from organic crystals such as anthracene. Solutions of hyperconjugated, heteronuclear compounds in aromatic hydrocarbon solvents are more efficient. A benzene solution containing 1% 2,5-diphenyloxazole and 0.1% 1,4-bis-2-(4-phenyl-5-phenyloxazole)-benzene is only one of the many recipes for liquid scintillation cocktails. When the radioactive sample is dissolved or homogeneously suspended in such liquid scintillators, the counting geometry approaches 4π. The term 4π refers to the number of steradians around a point source in space. There are 4π steradians in a sphere. Under such conditions, the detector essentially completely surrounds the source of radiation.

The scintillation process essentially converts the energy of the beta radiation to minute flashes of light that stimulate the photocathode of a photomultiplier tube and generate an electrical signal proportional to the energy of the beta radiation. Descriptive literature prepared by Packard BioScience (2001) and National Diagnostics (2002) provides detailed information on the preparation of samples for liquid scintillation counting, the liquid scintillation process, the detection of the scintillations, and the management of sources of error.

The liquid scintillation counter is shown schematically in Figure 27.

In theory, liquid scintillation should be ideally suited to the measurement of low-energy beta emitters such as those from ^{3}H and ^{14}C. However, thermionic emission at the photocathode of the photomultiplier tube

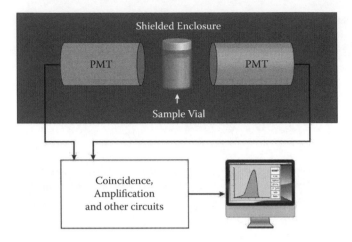

Figure 27 Liquid scintillation counter.

produces signals on the same order of magnitude as those from ^{3}H. This dark current makes a significant contribution to the apparent background activity. This contribution increases the uncertainties in the net count rate when the background is subtracted from the gross count rate. An example of this uncertainty in the net count rate is presented in Experiment 3.

The diagram of the liquid scintillation counter shown in Figure 27 contains a coincidence circuit to reduce the contribution of the dark current to the background. Scintillations from the sample bottle produce signals in both photomultiplier tubes simultaneously, while the random emission of electrons responsible for dark current occurs in only one or the other of the photomultiplier tubes at any give instant in time. The coincidence circuit will reject signals that do not arise simultaneously in both photomultiplier tubes. Such coincidence circuitry is able to reduce the apparent background from a few thousand to a few counts per minute.

Under ideal conditions, the liquid scintillation counter can approach 100% efficiency for measuring ^{14}C and ^{35}S and at least 50% efficiency for measuring ^{3}H. Unfortunately, conditions are rarely ideal. Frequently, components of the sample are able to absorb some of the scintillations before they reach the photocathode of the photomultiplier tube. The hemoglobin from hemolyzed red blood cells is notorious for its interference with the transmission of the scintillations in the scintillator (Moore, 1981). Similarly, fingerprints or condensate on the outer surface of the scintillation vial prevents transmission of scintillations to the photomultiplier tube. Collectively, the processes interfering with the transmission of scintillations are forms of quenching. Quench corrections are integral components of the proprietary programs that control liquid scintillation counters.

Typically, liquid samples are dissolved in scintillation cocktail and contained in small vials. These are loaded into a rack or cassette that is placed in the autosampler of the liquid scintillation counter. The lid is closed, and each sample is then automatically transported to a position between the photomultiplier tubes, counted, and returned to the cassette. Even though these operations are controlled by the software, this experiment should be supervised by an experienced operator.

Procedure

a. Prepare the following solutions by pipetting the volumes indicated into separate scintillation vials.
1. 4.5 ml of scintillation cocktail + 100 µl of tritiated water (5,000 Bq ^{3}H/ml)
2. 4.5 ml of scintillation cocktail + 100 µl of ^{14}C-ethanol (5,000 Bq ^{14}C/ml)
3. 4.5 ml of scintillation cocktail + 100 µl of tritiated water (5,000 Bq ^{3}H/ml) + 100 µl of ^{14}C-ethanol (5,000 Bq ^{14}C/ml)
4. 4.5 ml of scintillation cocktail + 100 µl of ^{14}C-ethanol (5,000 Bq ^{14}C/ml) + 20 µl of nitromethane
5. 4.5 ml of distilled water + 100 µl of ^{14}C-ethanol (5,000 Bq ^{14}C/ml)

b. Label the *tops* of the scintillation vials H, C, H-C, C-q, and C-w, respectively.
c. Load the scintillation vials into the cassette and place the cassette in the autosampler.
d. Under the supervision of the experienced operator, count the samples.

Experimental data

Observed activity of vial H	_____ cpm
Observed activity of vial C	_____ cpm
Observed activity of vial H-C	_____ cpm
Observed activity of vial C-q	_____ cpm
Observed activity of vial C-w	_____ cpm

Report

a. Calculate the expected activity of vial H in dpm, and determine the efficiency for the measurement of ^{3}H.
b. Calculate the expected activity of vial C in dpm, and determine the efficiency for the measurement of ^{14}C.

c. Calculate the expected activity of vial H-C in dpm, and comment on the efficiencies for the measurement of ^{3}H and ^{14}C.
d. Comment on the effect the addition of nitromethane had on the efficiency for the measurement of ^{14}C.
e. Comment on the effect the replacement of scintillation cocktail with water had on the efficiency for the measurement of ^{14}C.

Questions

1. Was quenching a significant issue in your experiment? Explain.
2. Why is it important to label the tops of the scintillation vials rather than the sides?
3. Would liquid scintillation be an efficient detection method for a radionuclide that emits only gamma radiation? How about a nuclide that emits alpha and gamma, but not beta? Briefly explain.
4. Why does detection take place in an enclosed, shielded space?
5. Why is detection efficiency so much lower for ^{3}H than for ^{14}C and ^{35}S? Qualitatively, what sort of efficiency would be expected for ^{32}P?

chapter twenty-two

Experiment 22 – separation by precipitation

The objectives of this experiment are to resolve the RaDEF secular equilibrium into its components by precipitation and to examine the changes in the activity of each component with respect to time.

In the distant past, selective precipitations have served as the basis for chemical separations. The classical "qual scheme" (Fresenius, 1866) and the tedium of gravimetric analysis were replaced by more modern techniques long ago. Nonetheless, there is some chemistry to be learned from these classical procedures.

^{238}U undergoes a series of sequential radioactive decays to stable ^{206}Pb. (See Bryan, 2009, Fig. 1.5, p. 11; Loveland et al., 2006, Fig. 3.11, p. 78.) The last three steps in this series of sequential radioactive decays are:

$$^{210}Pb \rightarrow {}^{210}Bi + \beta$$

$$^{210}Bi \rightarrow {}^{210}Po + \beta$$

$$^{210}Po \rightarrow {}^{206}Pb + \alpha$$

The three radionuclides, ^{210}Pb, ^{210}Bi, and ^{210}Po, constitute a secular equilibrium. (See Bryan, 2009, Section 2.6.1, pp. 27–29; Loveland et al., 2006, Section 3.3, pp. 73–77.) This secular equilibrium system can be resolved into its components by classical chemical procedures such as selective precipitations.

Južnič and Kobal (1986) have utilized selective precipitations for the radiochemical determination of ^{210}Po and ^{210}Pb in water, and Nevissi (1991) has measured ^{210}Pb, ^{210}Bi, and ^{210}Po in environmental samples after disequilibrium by precipitation.

Procedure

    a. Pipette the following into a small centrifuge tube: 1.0 ml of the bismuth carrier (50 mg/ml), 1.0 ml of the lead carrier (50 mg/ml), 0.1 µCi of RaDEF, and 2.0 ml of water.

    b. Add 1.0 ml of 10% H_2SO_4 solution to the centrifuge tube to precipitate $PbSO_4$.

c. Centrifuge and decant the supernate to a second centrifuge tube. Wash the precipitate with 1.0 ml of 10% H_2SO_4 solution, centrifuge, and decant the washings into the second centrifuge tube.

d. Wash the precipitate with 1.0 ml of water, centrifuge, and decant the washings to waste.

e. Transfer the $PbSO_4$ precipitate to a stainless steel planchette and dry it.

f. Render the contents of the second centrifuge tube basic with 10% NH_3 solution to precipitate the basic bismuth salt.

g. Centrifuge, and decant the supernate to waste. Wash the precipitate with water, centrifuge, and decant with washings to waste.

h. Transfer the basic bismuth salt to a stainless steel planchette and dry it.

i. Pipette 0.1 μCi of RaDEF into a stainless steel planchette and dry it.

j. Measure the activity of each planchette at daily intervals for 2 weeks.

Experimental data

Day	RaDEF Activity	$PbSO_4$ Activity	BiO(OH) Activity
0	_____ cpm	_____ cpm	_____ cpm
1	_____ cpm	_____ cpm	_____ cpm
2	_____ cpm	_____ cpm	_____ cpm
3	_____ cpm	_____ cpm	_____ cpm
4	_____ cpm	_____ cpm	_____ cpm
5	_____ cpm	_____ cpm	_____ cpm
6	_____ cpm	_____ cpm	_____ cpm
7	_____ cpm	_____ cpm	_____ cpm
8	_____ cpm	_____ cpm	_____ cpm
9	_____ cpm	_____ cpm	_____ cpm
10	_____ cpm	_____ cpm	_____ cpm
11	_____ cpm	_____ cpm	_____ cpm
12	_____ cpm	_____ cpm	_____ cpm
13	_____ cpm	_____ cpm	_____ cpm
14	_____ cpm	_____ cpm	_____ cpm

Report

a. Explain any changes in the activity of the isolated lead isotopes.
b. Explain any changes in the activity of the isolated bismuth isotopes.
c. Write balanced chemical equations to describe the separation steps.
d. For what reason(s) was it necessary to add bismuth carrier and lead carrier?
e. Which component is RaD?
f. Which component is RaE?
g. Which component is RaF?
h. What are the half-lives of each component?

Questions

1. What mass of ^{210}Bi is present in 10.0 mg of RaDEF?
2. The solubility product constant (K_{sp}—the solubility of relatively insoluble compounds) can easily be determined using radioactive tracers. A sample of $BaSO_4$ is placed in 250.0 ml of water at room temperature, and a small amount dissolves. The solid is filtered off and the activity of the aqueous solution is measured to be 4.29×10^{11} dpm. If the solid sample was 10^6:1 Ba:^{137m}Ba, what is the solubility of $BaSO_4$ in water at room temperature?

$$BaSO_4(s) \leftrightharpoons Ba^{2+}(aq) + SO_4^{2-}(aq) \qquad K_{sp} = [Ba^{2+}][SO_4^{2-}]$$

3. Polonium can be isolated from thorium solutions. What would be the activity due to polonium isolated from 1.0 ml of a 0.25 M $Th(NO_3)_4$ solution?

chapter twenty-three

Experiment 23 – chromatographic separation

The objectives of this experiment are to become familiar with the technique of paper chromatography, resolve the RaDEF secular equilibrium into its components by paper chromatography, and locate and identify each component on the chromatogram.

Paper chromatography is a powerful separation tool based on the components of mixtures undergoing many partitions between a stationary phase and a mobile phase. Each component of the mixture has characteristic affinities for the stationary phase and the mobile phase. Consequently, those with the greater affinities for the mobile phase will be more mobile, and those with the greater affinities for the stationary phase will be retarded in their movement. Each component of the mixture has a characteristic retardation factor, R_f. The R_f is the ratio distance traveled by the component to the distance traveled by the mobile phase. The RaDEF equilibrium system described in Experiment 22 can be resolved into its components by paper chromatography.

Procedure

    a. Mark a starting line 2-cm from the narrow edge of a 2 × 25-cm strip of chromatography paper.

    b. Pipette 0.02 μCi of RaDEF onto the center of this starting line.

    c. Allow the spot to dry.

    d. Suspend the strip of paper in the chromatography chamber as shown in Figure 28.

    e. Develop the chromatogram with 1 M HCl saturated 1-butanol.

    f. Remove the paper strip from the chromatography chamber when the mobile phase has traveled approximately 20-cm up the paper.

    g. Mark the leading edge of the mobile phase.

    h. Dry the chromatogram, and cut it into 5-mm segments along its length.

    i. Measure the activity of each segment, and retain those segments showing the highest activities for periodic measurements to determine half-lives using the procedures described in Experiment 11.

Data

Segment	Activity	Segment	Activity
0.0–0.5		10.0–10.5	
0.5–1.0		10.5–11.0	
1.0–1.5		11.0–11.5	
1.5–2.0		11.5–12.0	
2.0–2.5		12.0–12.5	
2.5–3.0		12.5–13.0	
3.0–3.5		13.0–13.5	
3.5–4.0		13.5–14.0	
4.0–4.5		14.0–14.5	
4.5–5.0		14.5–15.0	
5.0–5.5		15.0–15.5	
5.5–6.0		15.5–16.0	
6.0–6.5		16.0–16.5	
6.5–7.0		16.5–17.0	
7.0–7.5		17.0–17.5	
7.5–8.0		17.5–18.0	
8.0–8.5		18.0–18.5	
8.5–9.0		18.5–19.0	
9.0–9.5		19.0–19.5	
9.5–10.0		19.5–20.0	

Report

a. Using Excel™ or similar software, plot the observed radioactivity against the distance from the starting line.

b. Identify the segments or zones showing elevated radioactivity.

c. Calculate the retardation factors, R_f, for each zone.

d. Determine the half-lives of the radioactive species at each zone.

e. On the basis of half-life, identify the radiochemical species at each zone.

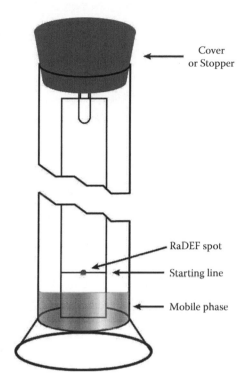

Figure 28 Paper chromatography of RaDEF.

Questions

1. What mass of ^{210}Pb is present in 0.02 µCi of RaDEF?
2. Can the R_f factors of the radioactive species be correlated with their relative affinities for the mobile and stationary phases?
3. Why isn't the RaDEF spot placed at the very bottom (edge) of the paper?
4. Why isn't the separation terminated after the leading edge of the mobile phase reaches the top of the chromatogram?

chapter twenty-four

Experiment 24 – random errors

The objectives of this experiment are to observe the random nature of radioactive decay and to evaluate these observations by statistical means.

It is important to distinguish between systematic and random errors. The latter follow the well-defined statistical laws.

Gaussian statistics can be applied to the random nature of radioactive decay. Such applications involve the calculation of the mean, N, and its standard deviation, σ. For the mean, $N = (1/N) \Sigma n_i$, and for the standard deviation, $\sigma = \sqrt{N}$.

Procedure

Make 20 one-minute counts on a source giving 2,000 to 5,000 cpm.

Experimental data

Measurement	Activity	Measurement	Activity
1	_____	11	_____
2	_____	12	_____
3	_____	13	_____
4	_____	14	_____
5	_____	15	_____
6	_____	16	_____
7	_____	17	_____
8	_____	18	_____
9	_____	19	_____
10	_____	20	_____

Report

a. Using the standard deviation calculator (mathisfun.com, 2009) or similar software, calculate the mean and the standard deviation for the data listed above.

Mean _____   σ _____

b. How many of the observations are within ±2 σ?

c. How many of the observations are expected to be within ±2 σ?

d. Do the data follow a normal distribution?

Questions

1. Could the ages of everyone present in the laboratory section fit a Gaussian distribution? Why or why not?

2. Do the data tabulated here fit a Gaussian distribution? Briefly justify your answer.

2,037	2,018	2,019	2,039	2,028
2,055	1,950	1,970	1,961	1,977
2,050	2,073	1,958	2,058	1,989
2,018	2,007	1,990	2,007	2,072

chapter twenty-five

Experiment 25 – duplicate samples

The objectives of this experiment are to develop laboratory skills in the manipulation of solutions containing radioactive material and to apply statistical theory to the interpretation of experimental results.

Radiations from solids, liquids, and gases can be detected and measured. Very often, samples are counted as dry solids to maintain a constant geometry and to minimize the potential for a spill that would contaminate the equipment and the working environment.

The difference between the activities measured for two samples of radioactive material may be due to systematic or random error. Statistical analysis of the experimental results can evaluate the probability of the latter.

The probability of observing a difference between two values can be estimated using the normal distribution, $P = 1/(\sigma\sqrt{2\pi})\,e^{-(n-\mu)2/2\sigma2}$. This probability function has been solved for many values of P, but the results are tabulated in several ways. The one-tailed, less than tables are most common. These tables present the probabilities of observing a difference less than or equal to the listed values. Because they are one-tailed, the listed values are doubled to obtain the probability.

Procedure

a. On the basis of the label information, calculate the volume of radioactive stock solution required to give 2,000 to 5,000 cpm assuming a 5% counting efficiency. (See the example on p. 20 of Bryan's (2009) text for help with this calculation.)

b. Using water instead of radioactive material, practice manipulation of the pipetting devices.

c. Label two planchettes as sample 1 and sample 2.

d. Pipette the calculated volume of stock solution into each planchette.

e. Dry the contents of the planchettes under the infrared lamp.

f. Count the samples for 1.0 and 5.0 minutes.

Experimental data

Isotope: _____   Activity: _____ μCi/ml   Volume: _____ μl

Expected activity _____ dpm   Observed activity: _____ cpm

	Counts in 1 minute	Counts in 5 minutes
Sample 1	_____	_____
Sample 2	_____	_____

Report

a. Calculate the relative error associated with each set of measurements as

$$\text{r.e.} = (r_1 - r_2) \div [\sqrt{(r_1/t_1 + r_2/t_2)}]$$

r.e. for 1-minute counts   _____
r.e. for 5-minute counts   _____

b. Using a standard normal distribution table (mathisfun.com, 2009), determine the probability of observing relative errors this large or smaller.

c. Determine the probability the samples are duplicates.

Questions

1. If the dead time loss was 2.5% at 4,000 cpm, would it be significant? In other words, what is the error associated with a single 1-minute count of 4,000 cpm?

2. Select some data from a previous experiment where the background rate was also measured and a background correction applied. Was the background correction statistically significant?

3. If there were a choice of collecting a single 1-minute count or a single 5-minute count, which should be chosen to obtain higher precision?

4. A sample gives a gross count of 912 for 60 minutes. The background on this counter was 895 for 60 minutes. Calculate the net cpm of the sample. Calculate the uncertainty in the net counts and cite the results to ±2σ. If either the sample or the background could be counted for 120 minutes, which should be chosen? Explain.

chapter twenty-six

Experiment 26 – measurement of neutron flux

The objectives of this experiment are to determine the half-life of ^{116m1}In and the neutron flux for a small plutonium-beryllium, PuBe, neutron source.

In nature, indium is present as ^{113}In (4.23%) and ^{115}In (95.77%). Consequently, several nuclear reactions take place when indium is bombarded with neutrons. These are summarized in Table 5.

In addition to nuclear reactors and particle accelerators, neutrons are produced in the reaction between beryllium and alpha radiation, $^9_4\text{Be} + ^4_2\text{He} \rightarrow ^{12}_6\text{C} + ^1_0\text{n}$. Plutonium, ^{239}Pu, decays by alpha emission and serves as the source of alpha radiation. Chatt and Katz (1980) have published a comprehensive description of neutron sources for activation analysis.

The intensity of the radioactivity induced in indium depends upon several parameters, $A = \sigma\varphi N(1 - e^{-\lambda t})$, where A is the intensity of the induced radioactivity in dps, σ is the cross section for the reaction in barns (b = 10^{-24}cm^2), N is the number of target nuclei, φ is the neutron flux in cm$^{-2} \cdot$ s^{-1}, λ is the decay constant, $0.693/t_{1/2}$ of the product nuclide, and t is the duration of the radioactivation expressed in the same time units as the half-life.

The intensity of neutron radiation is often the limiting factor in determining the sensitivity of neutron activation analysis. Knowledge of the intensity of neutron radiation or neutron flux is needed to optimize radioactivation.

Table 5 Some Nuclear Properties of Indium

Target	Product	Cross Section	Product Half-Life
^{113}In	^{114}In	4 b	72 s
^{113}In	^{114m}In	8 b	50 days
^{115}In	^{116}In	45 b	14 s
^{115}In	^{116m1}In	145 b	54 min
^{115}In	^{116m2}In	4 b	2 s

Procedure

a. Weigh an indium foil and secure it to a foil holder.
b. Note the time, insert the foil holder into the neutron source, and irradiate the foil for exactly 54 minutes.
c. Measure the background activity.
d. While the foil is being irradiated, determine the detector efficiency
   of a Geiger-Müller counter using a RaDEF reference source.
e. At the end of the 54-minute irradiation period, remove the foil holder
   from the neutron source, note the time, and mount the foil in a ring
   and disc under an aluminum cover.
f. Begin counting the foil exactly 5 minutes after its removal from the
   neutron source. Take 1-minute counts every 5 minutes for 1 hour.

Experimental data

a. Mass of indium foil _____ mg
b. Time in _____   Time out _____   Duration of activation _____ s
c. Background activity _____ cps
d. Activity of RaDEF reference source _____ µCi _____ dps
e. Measured activity of RaDEF source _____ cpm _____ cps
f. Measured activity of radioactivated indium foil:

Time	Activity	Time	Activity
0			
0 + 5	_____	0 + 35	_____
0 + 10	_____	0 + 40	_____
0 + 15	_____	0 + 45	_____
0 + 20	_____	0 + 50	_____
0 + 25	_____	0 + 55	_____
0 + 30	_____	0 + 60	_____

Report

a. Correct the observed count rates as necessary.
b. Using Microsoft Excel™ or similar software, plot the logarithm of
   the count rates against time, taking as zero time the time at which
   the foil was removed from the neutron source.
c. Extrapolate the decay curve to zero time.
d. Using the efficiency determined in step (d) of the procedure, calculate the activity of the indium in dps at zero time.

e. Calculate the neutron flux, φ, from $A = \sigma\varphi N(1 - e^{-\lambda t})$.

f. Determine the half-life of the activated indium, and compare the experimental value with those in the literature.

Questions

1. Explain why RaDEF is a suitable source for the determination of the detector efficiency in this case.

2. Why is the irradiated indium placed under aluminum for counting?

3. How important is it that the indium foil be pure?

4. How important is it that the foil be irradiated for exactly 54 minutes?

5. Write out all nuclear equations (reaction and decay) involving indium in this experiment.

6. Could this experiment be performed with yttrium instead of indium? Explain.

7. Calculate the percentage of indium in the sample converted to ^{116m1}In at time zero.

chapter twenty-seven

Experiment 27 – neutron activation analysis

The object of this experiment is to determine the concentration of manganese in a solution by neutron activation analysis.

Neutron activation is a two-step technique of elemental analysis in which a few atoms of the analyte are first rendered radioactive, and the intensity of the induced radioactivity, which is proportional to the number of analyte atoms in the sample, is then measured. Neutron activation analysis is a comparative technique; quantification is based on comparisons with the radioactivity induced in standards.

Manganese is monoisotopic, ^{55}Mn. Its tendency to undergo radioactivation is reflected in its cross-section, 13.3 b (1 b $= 1 \times 10^{-24}$ cm^2).

$$^{55}_{25}Mn + {}^{1}_{0}n \rightarrow {}^{56}_{25}Mn$$

The half-life of $^{56}_{25}$Mn is 2.57 hours. The decay scheme is complicated. It includes a 2.7 MeV β$^-$ and a 1.1 MeV γ.

Procedure

a. Prepare a reference series of MnSO$_4$ solutions by pipetting 0.0, 0.20, 0.40 0.60, 0.80, and 1.0 ml of 0.10 M MnSO$_4$ into separate plastic vials, and diluting the contents of each to 1.0 ml with water.

b. Pipette two 1.0 ml aliquots of the unknown manganese solution into separate vials, being certain to label them appropriately.

c. Place each vial into a polyethylene irradiation container, and position the containers in the neutron source. Irradiate the references and unknowns overnight.

d. At the end of the irradiation period, remove the irradiation containers singly, and note the corresponding times.

e. Remove the vial containing the manganese solution, and place it in the well of the NaI(Tl) scintillation detector. At a time corresponding to exactly 2 minutes after removal from the neutron source, make a 2-minute count. Repeat this procedure with the remaining manganese samples and standards.

Experimental data

Sample/standard	Activity	Manganese concentration
0.00 ml standard	_____ cpm	0.00 mg/ml
0.20 ml standard	_____ cpm	_____ mg/ml
0.40 ml standard	_____ cpm	_____ mg/ml
1.0 ml sample	_____ cmp	_____ mg/ml
0.60 ml standard	_____ cpm	_____ mg/ml
0.80 ml standard	_____ cpm	_____ mg/ml
1.0 ml sample	_____ cpm	_____ mg/ml
1.0 ml standard	_____ cpm	_____ mg/ml

Report

a. Using Microsoft Excel™ or similar software, construct a calibration curve by plotting the intensities of the induced radioactivities of the reference standards against their respective manganese concentrations.

b. Using the calibration curve data, determine the manganese concentration of the unknown solution as mg Mn/ml.

Questions

1. What is the product when $^{56}_{25}Mn$ undergoes decay by negatron emission?
2. If 100 dps corresponds to the detection limit for the determination of manganese by neutron activation analysis, what are the smallest masses of manganese measured:
   a. With a nuclear reactor where φ is 1×10^{12} cm$^{-2} \cdot$ s^{-1}?
   b. With a Cockroft-Walton generator where φ is 1×10^{8} cm$^{-2} \cdot$ s^{-1}?
   c. With a PuBe source where φ is 1×10^{3} cm$^{-2} \cdot$ s^{-1}?

chapter twenty-eight

Experiment 28 – hot atom chemistry

The objectives of this experiment are to explore the recoil of atoms upon neutron capture and to measure changes in chemical speciation upon neutron capture. While this phenomenon is known as the Szilard-Chalmers effect, Szilard and Chalmers (1934) used the term Fermi effect in their original report. Two typical examples are:

$$^{55}_{25}MnO_4^{1-} + {}^1_0n \rightarrow {}^{56}_{25}MnO_2 + H_2O$$

$$^{127}_{53}I\text{-}CH_2\text{-}CH_3 + {}^1_0n \rightarrow {}^{128}_{53}I^{1-} + CH_3\text{-}CH_3$$

Neutron capture is accompanied by the emission of a prompt gamma photon and the recoil of the radioactivated nuclide. Recoil energies are well in excess of typical bond energies. Consequently, bonds are broken, and the radioactivated nuclide often appears in a chemical form different from that of the original target nuclide. The process is described as the Szilard-Chalmers effect. This change in chemical form of the "hot" atom has been used to prepare high specific activity radiotracers for basic research and clinical applications.

$Na_2^{51}CrO_4$ is used for the *in vivo* measurement of blood volume and erythrocyte survival time. Preparation of this radiotracer by the chemical oxidation of radioactivated chromium metal yields a product of low specific activity because very few of the ^{50}Cr atoms in the metal actually undergo radioactivation. Radioactivation of the chromium in sodium chromate, however, produces trivalent 51-chromium, which is readily separated from the large excess of stable hexavalent chromium. After separation, the trivalent 51-chromium is oxidized to the hexavalent state and sterilized before it is used in clinical assessments. (See Section 2.3, pp. 22–23, of Bryan's (2009) text for a discussion of specific activity.)

Low Specific Activity

Radioactivate	$^{50}_{24}Cr + ^{1}_{0}n \rightarrow ^{51}_{24}Cr$ + large excess of stable chromium
Dissolve and oxidize	$^{51}_{24}Cr$ + large excess of stable chromium $\rightarrow ^{51}_{24}CrO_4^{2-}$ + large excess of stable chromium as chromate; hence, the radioactive chromium is of low specific activity

High Specific Activity

Radioactivate	$^{50}_{24}CrO_4^{2-} + ^{1}_{0}n \rightarrow ^{51}_{24}Cr^{3+}$ + large excess of stable chromium as chromate
Separate	Anion ion exchange chromatography or solvent extraction
Oxidize	$^{51}_{24}Cr^{3+} \rightarrow Na_2^{51}CrO_4$; hence, the radioactive chromium is of high specific activity

The cross-section for the reaction $^{127}I + ^{1}_{0}n \rightarrow ^{128}I$ is 5.5 barns, and the half-life of the ^{128}I is 25 minutes. The energy associated with the reaction $^{127}I + ^{1}_{0}n \rightarrow ^{128}I$, known as an (n,γ) reaction, is on the order of 8 MeV. This energy is emitted as a prompt gamma, and the ^{128}I recoils with energy well in excess of the I–C bond energy, ~5 eV. The energetic ^{128}I is the hot atom because it dissipates its energy as it travels through the ethyl iodide. Chemical bonds are broken in the surrounding ethyl iodide, and the fragments recombine to form both inorganic and organic species containing ^{128}I. These can be separated by extracting the irradiated ethyl iodide with aqueous sodium thiosulfate solution.

Procedure

a. Using a 100-ml separatory funnel, preclean 30-ml of ethyl iodide by extracting the inorganic impurities into 30-ml of an aqueous solution containing 2% sodium thiosulfate solution and 2% potassium iodide. Discard the aqueous phase.

b. Transfer the precleaned ethyl iodide to a 50-ml plastic bottle, and irradiate it in the neutron source for 1 hour.

c. Transfer the irradiated ethyl iodide to a 50-ml separatory funnel, and extract it with three separate 5.0-ml portions of an aqueous solution containing 2% sodium thiosulfate solution and 2% potassium iodide.

d. Combine the aqueous extracts.

e. If a well type sodium iodide scintillation detector is available, transfer 10 ml of the aqueous extracts and 10 ml of the nonaqueous phase to separate tubes, and measure the radioactivity of each phase. Repeat the measurements of radioactivity every 10 minutes for 1 hour.

f. If a well type sodium iodide scintillation detector is not available, transfer 1.0 ml aliquots of each phase to separate planchettes, dry them, and measure the radioactivity of each phase. Repeat the measurements of radioactivity every 10 minutes for 1 hour.

Experimental data

Time	Activity of aqueous phase	Activity of organic phase
0 min	_____ cpm	_____ cpm
0 + 10 min	_____ cpm	_____ cpm
0 + 20 min	_____ cpm	_____ cpm
0 + 30 min	_____ cpm	_____ cpm
0 + 40 min	_____ cpm	_____ cpm
0 + 50 min	_____ cpm	_____ cpm
0 + 60 min	_____ cpm	_____ cpm

Report

a. Calculate the half-life of the ^{128}I formed in this experiment, and compare the experimental result with the literature value.
b. Compare the radioactivities of the aqueous and organic phases, and offer explanations for these observations.

Questions

1. Calculate the total number of ^{127}I atoms converted to ^{128}I. How does this compare to the number of ^{127}I present prior to irradiation?
2. What is the advantage to using a well type detector in this experiment?
3. Write out balanced nuclear equations for the decay of ^{128}I. Does the fact that it is branched have a significant effect on the results of your experiment?

chapter twenty-nine

Experiment 29 – synthesis of ^{14}C aspirin

The objectives of this experiment are to synthesize and characterize acetyl ^{14}C-labeled aspirin.

The utilization of radiotracers is indispensable in many areas of scientific research, particularly in biochemical research. The elucidation of the photosynthetic cycle (Calvin and Benson, 1948) is only one example of how radiotracers have been employed. The radiotracer used here was $^{14}CO_2$, which is a readily synthesized, simple molecule. Research on metabolic pathways and pharmacokinetics, however, often requires the synthesis of more complicated radiotracers.

Rabinowitz et al. (1982) studied the dermal absorption of aspirin in dogs and humans using ^{14}C-labeled aspirin, and Hutt et al. (1982) studied the pharmacokinetics of aspirin in the human using carboxyl-^{14}C aspirin.

The synthesis of aspirin, as shown in Figure 29, is straightforward. The synthesis becomes more challenging when the ^{14}C must be in a specific location in the molecule. The synthesis shown in Figure 29 indicates acetyl-^{14}C aspirin is the reaction product.

Procedure

a. Accurately weigh 300 mg of salicylic acid into a 30-ml beaker.
b. Pipette 1.0 ml of ^{14}C-acetic anhydride into the beaker slowly while shaking the beaker gently. Add a drop of concentrated sulfuric acid directly to the contents of the beaker. Mix the contents of the beaker thoroughly, and heat the mixture in a boiling water bath for 15 minutes.
c. Cool the mixture in an ice bath. Stir the mixture with a glass rod until it thickens into a semisolid mass of crystals. Add 5.0 ml of water and stir until only a thin sludge remains.
d. Filter the suspension through a small Buchner funnel using a pipette to transfer the remaining material to the funnel. Rinse the crystals with ice-cold water, and transfer the washings to the funnel. Draw air through the funnel to dry the crystals.

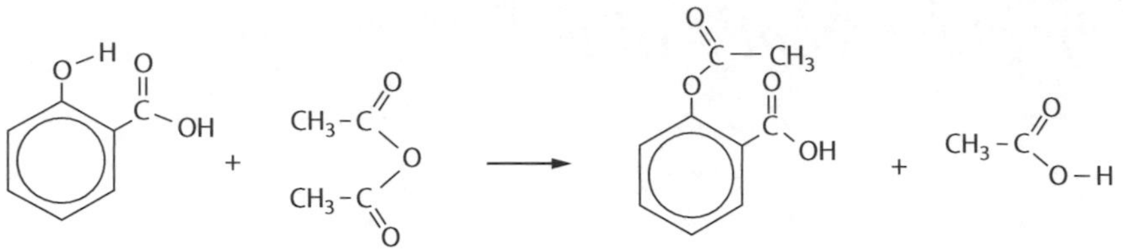

C$_7$H$_6$O$_3$	C$_4$H$_6$O$_3$	C$_9$H$_8$O$_4$	C$_2$H$_4$O$_2$
Salicylic acid	Acetic anhydride	Acetyl salicylic acid ASPIRIN	Acetic acid

Figure 29 Synthesis of aspirin.

  e. Transfer the crystals to a small flask, add 5 ml of toluene, and heat the suspension on a steam bath until the crystals dissolve. Cool the solution in an ice bath to recrystallize the product.
  f. Filter the recrystallized product through a preweighed Buchner funnel. Draw air through the crystals until they are dry. Reweigh the funnel to determine the chemical yield.
  g. Transfer the product to a planchette, and count it with a G-M counter. Determine the specific activity of the product.
  h. Determine the melting point of the product.
  i. If the instrumentation for infrared and nuclear magnetic resonance spectrometry is available, record the spectrum of the product for comparison to reference spectra.
  j. Deliver the product to the laboratory supervisor with a written statement of its mass and activity.

Experimental data

Mass of aspirin recovered	_____	mg
Activity of aspirin recovered	_____	cpm
Melting point of aspirin	_____	°C

Report

Prepare a statement containing the percent yield, purity, and specific activity of the acetyl-^{14}C aspirin prepared in this experiment. Include the infrared and nuclear magnetic resonance spectra if the instrumentation for recording them is available.

Questions

1. Which reactant is the limiting reagent in the synthesis of the ^{14}C-labeled aspirin?
2. Calculate the theoretical yield on the basis of the limiting reagent.
3. A student performing this experiment observes a lower than expected count rate. Give at least two reasons why a systematically low value might be observed.
4. Briefly outline how an animal study of the metabolic pathway(s) of aspirin might be performed using the material synthesized in this experiment.

chapter thirty

Experiment 30 – synthesis of ^{35}S-sulfanilamide

The objectives of this experiment are to synthesize and characterize ^{35}S-labeled sulfanilamide.

The first synthesis of sulfanilamide was reported in 1909. In 1935, sulfanilamide was found to inhibit the growth of bacteria such as staphylococcus. Sulfanilamide is one member in the family of sulfa drugs (sulfonamides) that interfere with metabolic processes of bacteria requiring para-amniobenzoic acid (PABA). Sulfanilamide competes with PABA, the substrate required by dihydropteroate synthetase for the synthesis of tetrahydrofolic acid. For humans, folic acid is a necessary component of the diet because humans are unable to synthesize folic acid. Hence, sulfanilamide is selectively toxic to staphylococcus. The similarities in the structures of sulfanilamide and PABA are shown in Figure 30.

Byrne et al. (1952) described the synthesis of several ^{35}S-labeled sulfa drugs, and Fingl et al. (1950) used ^{35}S-labeled N^4-acetyl-sulfanilamide to investigate the distribution of sulfanilamide in the rat after oral administration of the drug.

Procedure

a. Pipette 1.0 ml of carrier-tracer ^{35}S-sulfuric acid into a 30-ml beaker.
b. Cool the beaker and its contents in an ice water bath.
c. Slowly add 2.0 ml of acetic anhydride to the beaker while mixing thoroughly.
d. Add 1.0 g of acetanilide in small portions to the beaker while mixing thoroughly.
e. Heat the beaker and its contents on a steam bath in the hood for 30 minutes.
f. Cool the beaker and its contents in an ice bath, and add 5 ml of cold acetone to the beaker while mixing thoroughly.
g. Collect the crystals of N-acetylsulfanilic acid by filtration using a small Buchner funnel.
h. Wash the crystals on the filter with 2 ml of cold acetone, and dry them for 5 minutes 1 foot below a heat lamp while drawing air through the filter.

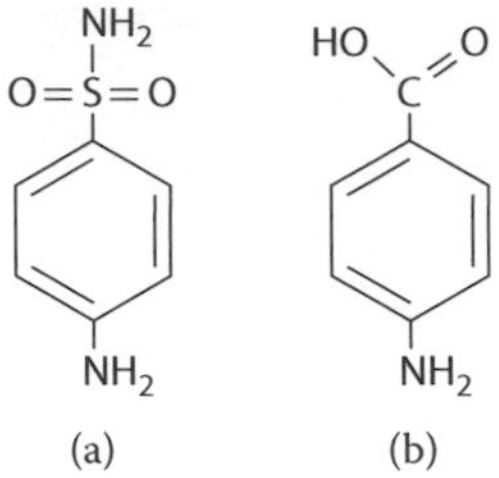

Figure 30 Structures for (a) sulfanilamide and (b) para-amniobenzoic acid (PABA).

i. Carefully transfer the crystals of N-acetylsulfanilic acid to a 30 ml beaker.

j. Cool the beaker and its contents in an ice water bath.

k. Slowly add 3.0 g of phosphorous pentachloride to the beaker while mixing thoroughly.

l. Cautiously heat the beaker and its contents on a steam bath until the contents become a homogeneous liquid.

m. Cool the beaker and its contents in an ice bath.

n. Add 5 g of crushed ice to the beaker to decompose the excess phosphorus pentachloride and precipitate the acid chloride of N-acetylsulfanilic acid.

o. Slowly add 5 ml of concentrated ammonia to the beaker while mixing thoroughly.

p. Heat the beaker and its contents on a steam bath in the hood for 30 minutes.

q. Cool the beaker and its contents to room temperature, and adjust the pH to 3 by the drop-wise addition of dilute sulfuric acid using the blue-to-yellow transition of bromphenol blue indicator to control the neutralization.

r. Collect the crystals of N-acetylsulfanilamide by filtration using a small Buchner funnel.

s. Wash the crystals of N-acetylsulfanilamide on the filter with 2 ml of cold water, and dry them for 5 minutes 1 foot below a heat lamp while drawing air through the filter.

t. Transfer the crystals of N-acetylsulfanilamide from the filter to a small test tube.

u. Add 2.5 ml of dilute (10%) hydrochloric acid to the test tube, and heat it with its contents 1 inch above a steam bath for 1 hour.

v. Cool the test tube and its contents in an ice water bath, and adjust the pH to 5 by the drop-wise addition of dilute (20%) sodium hydroxide solution using the yellow-to-blue transition of bromphenol blue indicator to control the neutralization. Add 5 ml of 10% sodium bicarbonate solution to the beaker.

w. Collect the crystals of sulfanilamide by filtration using a small Buchner funnel.

x. Wash the crystals of sulfanilamide on the filter with 2 ml of cold water, and dry them for 5 minutes 1 foot below a heat lamp while drawing air through the filter.

y. Transfer the filter with the crystals of sulfanilamide to a planchette, and measure the radioactivity.

z. Determine the mass of the planchette with the filter and the crystals of sulfanilamide. Determine the mass of another planchette containing a blank filter.

Experimental data

Activity of sulfanilamide   _____ cpm
Mass of sulfanilamide   _____ mg

Report

a. Using structural formulas, write the chemical equations for each chemical change in the synthesis of sulfanilamide from acetanilide.

b. Calculate the chemical yield of sulfanilamide.

c. Calculate the specific activity of the sulfanilamide.

Questions

1. Which reactant is the limiting reagent in the synthesis of the ^{35}S-labeled sulfanilamide?

2. Calculate the theoretical yield on the basis of the limiting reagent.

3. Calculate the percent yield of the ^{35}S-labeled sulfanilamide.

chapter thirty-one

Experiment 31 – radiological monitoring

At the end of the laboratory session, the workers should be monitored for both exposure to radiation and for contamination with radioactive materials. The workstation should be monitored for contamination with radioactive materials. Assessments of worker exposure and contamination are made by recording pocket dosimeter readings at the beginning and at the end of the laboratory session and by sweeping over the body surfaces, including soles of the shoes, with a handheld survey meter. After all radioactive sources have been secured, the workstation is monitored for contamination using a handheld survey meter. In addition, wet and dry wipes of the working surfaces are measured at one of the counting stations in the laboratory.

Procedure

a. At the beginning of the laboratory session, adjust a pocket dosimeter to near zero, read the dosimeter, and record the reading.
b. At the end of the laboratory period, read the dosimeter and record the reading.
c. Using a handheld survey meter, slowly sweep from head to toe the front and back body surfaces of a coworker, paying particular attention to the hands and shoes. Record the readings at 25-cm intervals.
d. Make a diagram of the laboratory.
e. Using a handheld survey meter, slowly sweep across the working surfaces of the laboratory.
f. Being certain to wear gloves, wipe 1-m² areas of the laboratory working surfaces with 5-cm diameter circles of dry filter paper using an S-shaped motion.
g. Being certain to wear gloves, wipe 1-m² areas of the laboratory working surfaces with 5-cm diameter circles of moistened filter paper using an S-shaped motion.
h. Being certain to wear gloves, wipe randomly selected 1-m² areas of the floor with 5-cm diameter circles of moistened filter paper using an S-shaped motion.

i. Place the filter papers used for the wipe tests as well as an unused filter paper on pieces of plastic sheet and measure the radioactivity using a Geiger-Müller detector and, if available, a sodium iodide scintillation detector.

Experimental data

a. Initial reading of pocket dosimeter _____ mSv
b. Final reading of pocket dosimeter _____ mSv
c. Background count rate of handheld survey meter _____ cpm
d. Background count rate of handheld survey meter _____ cpm
e. Background count rate of handheld survey meter _____ cpm
f. Record the results from the measurements made by the coworker on your body surfaces with the survey meter.
g. Record the results from the survey meter measurements on the diagram of the laboratory:

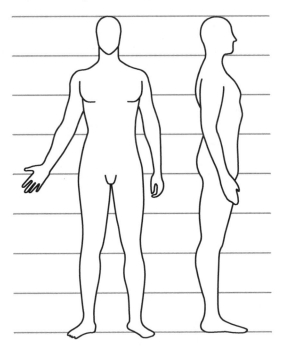

Figure 31 This diagram is used to record date for body contamination.

h. Record the results from the measurements on wipes of the working surfaces:

Location	Dry wipe G-M	Wet wipe G-M	Dry wipe NaI	Wet wipe NaI
Unused	_____	_____	_____	_____
1	_____	_____	_____	_____
2	_____	_____	_____	_____
3	_____	_____	_____	_____
4	_____	_____	_____	_____
5	_____	_____	_____	_____
Floor	_____	_____	_____	_____
Floor	_____	_____	_____	_____
Floor	_____	_____	_____	_____

Report

a. Exposure dose from dosimeter readings _____ mSv

b. Contamination of personnel

_____ location   _____ cpm or mSv

_____ location   _____ cpm or mSv

_____ location   _____ cpm or mSv

c. Contamination of working environment

_____ location   _____ cpm G-M   _____ cpm NaI

_____ location   _____ cpm G-M   _____ cpm NaI

_____ location   _____ cpm G-M   _____ cpm NaI

chapter thirty-two

Experiment 32 – determination of an unknown

The purpose of this experiment is to determine the identity of an element in an unknown sample. All of the unknowns contain one element that will activate by neutron radiation forming at least one radioactive nuclide, with a half-life between 20 seconds and 5 days. The unknowns either are in the elemental form or contain other elements that will not form significant quantities of radioactive products in the howitzer. While the unknowns have only one element that will become active, more than one radionuclide might be formed, making multicomponent decay a possibility.

A convenient way to guess the identity of the radionuclide produced is "The Lund/LBNL Nuclear Data Search" (http://nucleardata.nuclear. lu.se/nucleardata/toi/). Click on the "Radiation search" link. Enter the energy of the most prominent gamma peak and specify an error range. Comparison of actual and calculated energy values for peaks with similar energies from known radionuclides can give an idea of how much error to expect. Also enter a range for half-life of the activated sample. The previous experiments involving half-life determinations can serve as a guide for deciding on an appropriate range.

Click on the search button, and take a look at the possible nuclides. A long list may be generated. Screen out most of them by scanning down the relative intensity column. If the relative intensity is less than 1, that nuclide can probably be discarded. Further screening can be accomplished by discarding nuclides that decay via α, β^+, or electron capture (EC), because neutron capture by a stable nuclide will *most likely* produce a negatron-emitting nuclide. Remember how the radionuclide was produced!

Procedure

a. Irradiate the unknown sample in the neutron source. The optimal amount of time for this step will vary greatly, depending on the half-life of the radionuclide produced and the probability of a neutron reaction. A good rule of thumb is to irradiate ~2 to 3 times the half-life of the radionuclide produced. Some trial and error will be necessary to determine this.

b. Record the gamma spectrum and determine the half-life (or half-lives).

Report

a. Determine the identity of the unknown element.
b. Explain everything observed in the gamma spectrum.
c. Compare the experimentally determined half-life with the literature value.

Bibliography

Auger, P. 1925. Sur l'effet photoélectrique composé. *J. Phys. Radium*, 6, 205–8.

Becquerel, E. 1896. Sur les radiations emises par phosphorescence. *Comp. R. Hebd. Séan. Acad. Sci.*, 122, 420–21.

Boyarshinov, L. M. 1967. Dependence of the intensity of back-scattered beta particles on the atomic number of the target material with filtration of the back-scattered radiation. *Izvestiya. Vuz. Fizika*, 10, 152–55.

Bryan, J. C. 2009. *Introduction to nuclear science*. CRC Press, Boca Raton, FL, pp. 6–8, 88–90, 107–15.

Byrne, P. J., Alberts, A. A., and Christian, J. E. 1952. The synthesis of certain sulfur-35 labeled sulfa drugs. *J. Am. Pharm. Assoc.*, 42, 77–79.

Calvin, M., and Benson, A. A. 1948. The path of carbon in photosynthesis. *Science*, 107, 476.

Carswell, D. J. 1967. *Introduction to nuclear chemistry*. Elsevier, Amsterdam, p. 260.

Chase, G. D., and Rabinowitz, J. L. 1962. *Principles of radioisotope methodology*. Bugress, Minneapolis, pp. 114, 129.

Chatt, A., and Katz, S. A. 1980. Neutron sources for activation analysis of geological materials. In *Short course in activation analysis in the geosciences*, ed. G. K. Muecke. Minerological Society of Canada, Halifax, pp. 43–72.

Choppin, G. R., and Rydberg, J. 1980. *Nuclear chemistry theory and applications*. Pergamon Press, Oxford, p. 379.

Dickstein, P., Broshi, L., and Haquin, G. 2008. Empirical compensation for the precipitation effect in the measurement of gross alpha content in drinking water. *Health Physics*, 95 (Suppl. 5), S167.

Ehmann, W. D., and Vance, D. E. 1991. *Radiochemistry and nuclear methods of analysis*. Wiley, New York.

Feather, N. 1938. Further possibilities for the absorption method of investigating the primary β particles from radioactive substances. *Proc. Cambr. Philos. Soc.*, 34, 599–611.

Fingl, E. G., Christian, J. J., and Edwards, L. D. 1950. Tissue distribution of S^{35} in the rat following oral administration of labeled sulfanilamide. *J. Am. Pharm. Assoc.*, 39, 693–95.

Fisher, K. A. 1982. Monolayer freeze-fracture autoradiography: Quantitative analysis of the transmembrane distribution of radioiodinated concanavalin A. *J. Cell Biol.*, 93, 155–163.

Friedlander, G., Kennedy, J. W., Macias, E. S., and Miller, J. M. 1981. *Nuclear and radiochemistry*. 3rd ed. Wiley-Interscience, New York, 1981, pp. 361, 606–50.

Fresenius, C. R. 1866. *Manual of qualitative analysis*. Wiley, New York.

Geiger, H., and Nuttall, J. M. 1911. The ranges of the α-particles from various radioactive substances and a relation between range and the period of transformation. *Phil. Mag.*, 22, 613–16.

Goeppert-Mayer, M. 1948. On closed shells in nuclei. *Phys. Rev.*, 235–39.

Goeppert-Mayer, M. 1950a. Nuclear configurations in the spin-orbit coupling model. I. *Empirical Evid.*, 78, 16–21.

Goeppert-Mayer, M. 1950b. Nuclear configurations in the spin-orbit coupling model. II. *Theor. Consid.* 78, 22–23.

Gora, E. K., and Hickey, F. C. 1954. Self-absorption correction in carbon-14 counting. *Anal. Chem.*, 26, 1158–61.

Hendler, R. W. 1959. Self-absorption for carbon-14: A new treatment yields a correction factor that is linearly related to the thickness of the sample. *Science*, 130, 772–77.

Hutt, A. J., Caldwell, J., and Smith, R. L. 1982. The metabolism of [carbonyl-^{14}C] aspirin in man. *Xenobiotica*, 12, 601–10.

Johnson, J. C., Langhorst, S. M., Koyalka, S. K., Volkert, W. A., and Ketring, A. R. 1992. Calculation of radiation dose at a bone-to-marrow interface using Monte Carlo modeling techniques (EGS4). *J. Nucl. Med.*, 33, 623–28.

Južnič, K., and Kobal, I. 1986. Radiochemical determination of ^{210}Po and ^{210}Pb in water. *J. Radioanal. Nucl. Chem.*, 102, 493–98.

Katz, L., and Penfold, A. S. 1952. Range-energy relation for electrons and the determination of beta-ray end-point energies by absorption. *Rev. Mod. Phys.*, 24, 28–44.

Lambie, D. A. 1964. *Techniques for the use of radioisotopes in analysis*. Van Nostrand, London, pp. 112–13.

Lederer, C. M., Hollander, J. M., and Perlman, I. 1967. *Table of isotopes*. 6th ed. Wiley, New York.

Lee, S. H., Gardner, R. P., and Jae, M. 2004. Determination of dead times in a recently introduced hybrid G-M counter dead time model. *J. Nucl. Sci. Technol.*, 4, 156–59.

Libby, W. F. 1955. *Radiocarbon dating*. 2nd ed. University of Chicago Press, Chicago, p. 15.

Liester, K. H. 2001. *Nuclear and radiochemistry fundamentals and applications*. Wiley, Berlin.

Loveland, W., Morrissey, D. J., and Seaborg, G. T. 2006. *Modern nuclear chemistry*. Wiley, Hoboken, NJ, 78, 520, 573.

mathisfun.com. 2009. Standard deviation calculator.

Meitner, L. 1923. Das beta-strahlenspektrum von UX1 and seine Deutung. *A. Phys.*, 17, 54–66.

Moore, P. A. 1981. Preparation of whole blood for liquid scintillation counting. *Clin, Chem.*, 27, 609–11.

National Diagnostics. 2002. Principles and applications of liquid scintillation counting, a primer of orientation. National Diagnostics Laboratory Staff, www2.fpm. wisc.edu/safety/Radiation/docs/lsc_guide.pdf.

Nevissi, A. E. 1991. Measurement of lead-210, bismuth-210 and polonium-210 in environmental samples. *J. Radioanal. Nucl. Chem.*, 148, 121–31.

NIST. 1996. http://physics.nist.gov/PhysRefData/XrayMassCoef/cover/html.

NRC. 2008. NRC Regulations, Title 10, Code of Federal Regulations. GPO, Washington, DC.

Packard BioScience. 2001. Liquid scintillation counting application. Note LSC-007. Jock Thompson, www.americanlaboratorytraining.com/product-details-Packard-BioScience-Company-9069.html.

Preuss, L. E. 1952. Constant geometry for dead-time determinations. *Nucleonics*, 10, 62.

Rabinowitz, J. L., Feldman, E. S., Weinbaum, A., and Schumaker, H. R. 1982. Comparative tissue absorption of oral ^{14}C-aspirin and topical triethanolamine ^{14}C-salicylate in human and canine knee joints. *J. Clin. Pharmacol.*, 22, 42–48.

Röntgen, W. K. 1895. Über eine nue Art von Strahlen. Sitzungsberichte der Physikalisch-Medicinischen Gesellschaft zu Würtzburg, December 28.

Ruberu, S. R., Liu, Y.-G., and Perera, S. K. 2008. An improved liquid scintillation counting method for the determination of gross alpha activity in ground water samples. *Health Phys.*, 95, 397–406.

Scarr, E., Beneyto, M., Meador-Woodruff, J., and Dean, B. 2005. Cortical glutamatergic markers in schizophrenia. *Neuropsychopharmacology*, 30, 1521–31.

Schiff, L. I. 1936. Statistical analysis of counter data. *Phys. Rev.*, 50, 88–96.

Soller, M. S., Cirignano, L., Lieberman, P., and Squillante, M. R. 1990. A system for precise determination of effective atomic number. *IEEE Trans. Nucl. Sci.*, 37, 230–32.

Spatz, W. D. B. 1943. The factors influencing the plateau characteristics of self-quenching Geiger-Mueller counters. *Phys. Rev.*, 64, 236–40.

Szilard, L., and Chalmers, T. A. 1934. Chemical separation of the radioactive element from its bombarded isotope in the Fermi effect. *Nature*, 134, 462.

Wittwer, S. H., and Lundell, W. S. 1951. Autoradiography as an aid in determining the gross absorption and utilization of foliar applied nutrients. *J. Mich. Agric. Station*, 1233, 792–97.

Wood, K. G. 1971. Self-absorption corrections for the ^{14}C method with $BaCO_3$ for measurements of primary productivity. *Ecology*, 52, 491–98.

Appendix 1: some constants and factors

Constants

Avogadro's number	6.022×10^{23} mol^{-1}
Boltzman's constant	1.3807×10^{-23} J $\cdot$ K^{-1}
Planck's constant	6.626×10^{-34} J $\cdot$ s
Speed of light (vac.)	2.998×10^8 m $\cdot$ s^{-1}

Energy

1 u = 931.5 MeV
1 MeV = 1.602×10^{-13} J

Mass

Alpha	4.00150618 u
Electron	0.00054858 u
Neutron	1.00866492 u
Proton	1.00727647 u

1 u = 1.66054×10^{-24} g

Radioactivity

1 Ci = 3.7×10^{10} s^{-1}
1 Bq = 1 s^{-1}

Appendix 2: some nuclides

Element	Nuclide	Atomic Mass[a]	Abundance[b]	Half-Life	Radiations, Energies[c]
Hydrogen, Z = 1	^{1}H	1.007825	99.985	Stable	
	^{2}H	2.0140	0.015	Stable	
	^{3}H	3.01605	Trace	12.262 years	β^- 0.01861
Helium, Z = 2	^{3}He	3.10603	Trace	Stable	
	^{4}He	4.00260	99.999862	Stable	
Lithium, Z = 3	^{6}Li	6.01512	7.5	Stable	
	^{7}Li	7.01600	92.5	Stable	
Beryllium, Z = 4	^{7}Be	7.0169	None	53.37 days	EC; γ 0.477
	^{9}Be	9.01218	100	Stable	
	^{10}Be	10.0135	None	2.5×10^6 years	β^- 0.555
Boron, Z = 5	^{10}B	10.0129	20.0	Stable	
	^{11}B	11.00931	80.0	Stable	
Carbon, Z = 6	^{11}C	11.01143	None	20.3 min	EC; β^+ 0.97
	^{12}C	12.00000	98.90	Stable	
	^{13}C	13.00335	1.10	Stable	
	^{14}C	14.0032	Trace	5,730 years	β^- 0.156
Nitrogen, Z = 7	^{13}N	13.00574	None	9.965 min	β^+ 1.190
	^{14}N	14.00307	99.63	Stable	
	^{15}N	15.00011	0.37	Stable	

(continued)

127

Element	Nuclide	Atomic Mass[a]	Abundance[b]	Half-Life	Radiations, Energies[c]
Oxygen, Z = 8	^{15}O	15.0031	None	2.04 min	β^+ 1.74
	^{16}O	15.99491	99.762	Stable	
	^{17}O	16.9991	0.038	Stable	
	^{18}O	17.9992	0.200	Stable	
Fluorine, Z = 9	^{18}F	18.0009	None	109.7 min	β^+ 0.6335
	^{19}F	18.9984	100	Stable	
Neon, Z = 10	^{20}Ne	19.99244	90.51	Stable	
	^{21}Ne	20.99395	0.27	Stable	
	^{22}Ne	21.99138	9.22	Stable	
Sodium, Z = 11	^{22}Na	21.9944	None	2.60 years	β^+ 0.546; γ 1.275
	^{23}Na	22.9898	100	Stable	
	^{24}Na	23.99096	None	15.0 h	β^- 1.369; γ 1.38, 2.75
Magnesium, Z = 24	^{24}Mg	23.98504	78.99	Stable	
	^{25}Mg	24.98584	10.00	Stable	
	^{26}Mg	25.98256	11.01	Stable	
	^{27}Mg	26.98434	None	9.46 min	β^- 1.75: γ 0.84, 1.013
	^{28}Mg	27.98388	None	20.9 h	β^- 0.46; γ 0.95, 1.35
Aluminum, Z = 13	^{26}Al	25.98689	None	7.4×10^5 years	β^+ 1.17; γ 1.81
	^{27}Al	26.98154	100	Stable	
	^{28}Al	27.98191	None	2.24 min	β 2.85; γ 1.780
Silicon, Z = 14	^{28}Si	27.97693	92.23	Stable	
	^{29}Si	28.97649	4.67	Stable	
	^{30}Si	29.97376	3.10	Stable	
	^{31}Si	30.975362	None	2.62 h	β^- 1.48
	^{32}Si	31.9740	None	101 years	β^- 0.224
Phosphorus, Z = 15	^{31}P	30.97376	100	Stable	
	^{32}P	31.97391	None	14.3 days	β^- 1.71
	^{33}P	32.9717	None	25 days	β^- 0.248
Sulfur, Z = 16	^{32}S	31.97207	95.02	Stable	
	^{33}S	32.97146	0.75	Stable	
	^{34}S	33.96786	4.21	Stable	
	^{35}S	34.9690	None	87.9 days	β^- 0.167
	^{36}S	35.96709	0.02	Stable	
Chlorine, Z = 17	^{35}Cl	34.96885	75.77	Stable	
	^{36}Cl	35.9797	None	3.01×10^5 years	β^- 0.714

Element	Nuclide	Atomic Mass[a]	Abundance[b]	Half-Life	Radiations, Energies[c]
	^{37}Cl	36.9658	24.23	Stable	
	^{38}Cl	37.96801	None	37.2 min	β^- 4.91; γ 1.60, 2.17
Argon, Z = 18	^{36}Ar	35.96755	0.337	Stable	
	^{37}Ar	36.966776	None	35.0 days	EC
	^{38}Ar	37.96272	0.063	Stable	
	^{39}Ar	38,962314	None	268 years	β^- 0.565
	^{40}Ar	39.9624	99.600	Stable	
	^{41}Ar	40.964501	None	1.82 h	β^- 1.198; γ 1.293
Potassium, Z = 19	^{39}K	38.96371	93.2581	Stable	
	^{40}K	39.974	0.0117	1.28×10^9 years	EC; β^- 1.324; γ 1.460
	^{41}K	40.962	6.7302	Stable	
	^{42}K	41.963	None	12.4 days	β^- 3.52; γ 1.524
Calcium, Z = 20	^{40}Ca	39.96259	96.946	Stable	
	^{41}Ca	40.96227	None	1.02×10^5 years	EC
	^{42}Ca	41.95863	0.647	Stable	
	^{43}Ca	42.95878	0.135	Stable	
	^{44}Ca	43.95549	2.086	Stable	
	^{45}Ca	44.956	None	165 days	β^- 0.252
	^{46}Ca	45.9537	0.004	Stable	
	^{47}Ca	46.954	None	4.53 days	β^- 1.98; γ 1.308
	^{48}Ca	47.9524	0.187	Stable	
	^{49}Ca	48.955673	None	8.72 min	β^- 1.95
Scandium, Z = 21	^{44}Sc	43.959404	None	3.93 h	β^+ 1.47; γ 1.359
	^{45}Sc	44.95592	100	Stable	
	^{46}Sc	45.955	None	83.81 days	β^- 0.357; γ 0.889, 1.12
	^{47}Sc	46.955170	None	3.43 min	β^- 2.01
Titanium, Z = 22	^{44}Ti	43.9597	None	67 years	EC
	^{45}Ti	44.958124	None	3.078 min	EC; β^+ 2.07; γ 1.66
	^{46}Ti	45.9526294	8.25	Stable	
	^{47}Ti	46.9517640	7.44	Stable	
	^{48}Ti	47.9479473	73.72	Stable	
	^{49}Ti	48.9478711	5.41	Stable	
	^{50}Ti	49.9447921	5.81	Stable	

(continued)

Element	Nuclide	Atomic Mass[a]	Abundance[b]	Half-Life	Radiations, Energies[c]
	[51]Ti	50.946616	None	5.71 min	β⁻ 2.14; γ 0.320
Vanadium, Z = 23	[48]V	47.952	None	16.0 days	
	[49]V	48.948	None	330 days	
	[50]V	49.9472	0.250	1.4 × 10[17] years	EC; γ 1.554, and β⁺ 0.254; γ 0.738
	[51]V	50.9440	99.750	Stable	
	[52]V	51.94478	None	3.76 min	β⁻ 2.54; γ 1.434
Chromium, Z = 24	[49]Cr	48.951341	None	42.3 min	EC
	[50]Cr	49.9461	4.35	Stable	
	[51]Cr	50.945	None	27.8 days	EC; γ 0.320
	[52]Cr	51.9405	83.97	Stable	
	[53]Cr	52.9407	9.50	Stable	
	[54]Cr	53.9389	2.36	Stable	
	[55]Cr	54.940844	None	3.50 min	β⁻ 2.494
Manganese, Z = 25	[53]Mn	52.941294	None	3.7 × 10[6] years	EC
	[54]Mn	53.940363	None	312 days	EC
	[55]Mn	54.9380411	100	Stable	
	[56]Mn	55.948909	None	2.58 h	β⁻ 1.216; γ 0.847
Iron, Z = 26	[52]Fe	51.94812	None	8.28 h	EC; γ 0.169; β⁻ 2.494
	[53]Fe	52.945312	None	8.51 min	EC; β⁺ 3.82, 3.32
	[54]Fe	53.9396127	5.85	Stable	
	[55]Fe	54.938298	2.73	Stable	
	[56]Fe	55.9394393	91.75	Stable	
	[57]Fe	56.9353958	2.12	Stable	
	[58]Fe	57.9332773	0.28	Stable	
	[59]Fe	58.934880	None	44.5 days	β⁻ 0.466; γ 1.099
	[60]Fe	59.9340769	None	1.5 10[6] years	β⁻ 0.181
Cobalt, Z = 27	[57]Co	56.936296	None	272 days	EC; γ 0.122
	[58]Co	57.935757	None	70.9 days	EC; γ 0.610
	[59]Co	58.9331976	100	Stable	
	[60]Co	59.93382	None	5.27 years	β⁻ 0.318; γ 1.173, 1.33
Nickel, Z = 28	[56]Ni	55.94214	None	6.08 days	EC

Element	Nuclide	Atomic Mass[a]	Abundance[b]	Half-Life	Radiations, Energies[c]
	[57]Ni	56.939800	None	35.6 h	EC
	[58]Ni	57.9353462	68.1	Stable	
	[59]Ni	58.934351	None	7.6×10^4 years	EC
	[60]Ni	59.9307884	26.2	Stable	
	[61]Ni	60.9310579	1.1	Stable	
	[62]Ni	61.9283461	3.6	Stable	
	[63]Ni	62.929673	None	100 years	β^- 0.067
	[64]Ni	63.9279679	0.9	Stable	
Copper, Z = 29	[63]Cu	62.9295989	69.2	Stable	
	[64]Cu	63.939768	None	12.7 h	EC; γ 1.346; β^+ 1.673
	[65]Cu	64.9277929	30.8	Stable	
	[66]Cu	65.928873	None	5.12 min	β^- 2.630; γ 1.039
Zinc, Z = 30	[64]Zn	63.9291448	48.6	Stable (?)	
	[65]Zn	64.92645	None	243 days	EC
	[66]Zn	65.9260347	27.9	Stable	
	[67]Zn	66.9271291	4.1	Stable	
	[68]Zn	67.9248459	18.8	Stable	
	[69m]Zn	68.926553	None	13.9 h	IT; γ 0.439
	[69]Zn	68.926553	None	56 min	β^- 0.905
	[70]Zn	69.925325	0.6	Stable	
Gallium, Z = 31	[67]Ga	66.928206	None	3.26 days	EC
	[68]Ga	67.927983	None	1.13 h	EC; β^+ 2.921
	[69]Ga	68.925580	60.2	Stable	
	[70]Ga	69.926027	None	21.1 min	β^- 1.62
	[71]Ga	70.9247005	39.9	Stable	
	[72]Ga	71.926372	None	14.1 h	β^- many; γ many
	[73]Ga	72.92517	None	74.9 h	β^- 1.224; γ 0.326
Germanium, Z = 32	[68]Ge	67.92810	None	270 days	EC
	[69]Ge	68.927973	None	1.63 days	EC; β^+ 1.203; γ 0.871
	[70]Ge	69.9242497	20.8	Stable	
	[71]Ge	70.924954	None	11.2 days	EC
	[72]Ge	71.9220789	27.5	Stable	
	[73]Ge	72.9234626	7.7	Stable	
	[74]Ge	73.211774	36.3	Stable	
	[75]Ge	74.922860	None	1.38 h	β^- 1.188

(continued)

Element	Nuclide	Atomic Mass[a]	Abundance[b]	Half-Life	Radiations, Energies[c]
	^{76}Ge	75.9214016	7.6	Stable	
	^{77}Ge	76.923549	None	11.3 h	β⁻ many; γ many
Arsenic, Z = 33	^{73}As	72.923825	None	80.3 days	EC
	^{74}As	73,923829	None	17.8 days	EC; β⁺ 1.962; γ 0.596; β⁻ 1.35, 0.717; γ 0.635
	^{75}As	74.9215942	100	Stable	
	^{76}As	75.922394	None	26.3 h	β⁻ many; γ many
Selenium, Z = 34	^{74}Se	73.9224746	0.9	Stable	
	^{75}Se	74.922524	None	120 days	EC
	^{76}Se	75.9192120	9.4	Stable	
	^{77}Se	76.9199125	7.6	Stable	
	^{78}Se	77.9173076	23.8	Stable	
	^{79}Se	78.918500	None	6.5×10^4 years	β⁻ 0.160
	^{80}Se	79.9165196	49.6	Stable	
	^{81}Se	80.917993	None	18.5 min	β⁻ 1.580
	^{82}Se	81.9166978	0.7	Stable	
	^{83}Se	82.919119	None	22.3 min	β⁻ many; γ many
Bromine, Z = 35	^{77}Br	76.921380	None	2.38 days	EC; γ 0.239
	^{78}Br	77.921146	None	6.46 min	β⁺ 2.552; γ 0.613
	^{79}Br	78.9183361	50.7	Stable	
	^{80}Br	79.9118530	None	17.7 min	EC
	^{81}Br	80.916289	49.3	Stable	
	^{82}Br	81.916805	None	1.47 days	β⁻ 0.444, 0.264; γ 0.776, 0.698, 0.619
Krypton, Z = 36	^{78}Kr	77.920396	0.4	Stable	
	^{79}Kr	78.920083	None	1.5 days	EC
	^{80}Kr	79.916380	2.3	Stable	
	^{81}Kr	80.916593	None	2.3×10^5 years	EC
	^{82}Kr	81.913482	11.6	Stable	
	^{83}Kr	82.914135	11.5	Stable	
	^{84}Kr	83.911507	57.0	Stable	
	^{85}Kr	84.912530	None	10.7 years	β⁻ 0.687; γ 0.514

Element	Nuclide	Atomic Mass[a]	Abundance[b]	Half-Life	Radiations, Energies[c]
	^{86}Kr	85.910616	17.3	Stable	
	^{87}Kr	86.913359	None	1.3 h	β⁻ many; γ many
Rubidium, Z = 37	^{83}Rb	82.91511	None	86.2 days	EC; γ 0.520
	^{84}Rb	83.91487	None	32.9 days	EC
	^{85}Rb	84.911794	72.2	Stable	
	^{86}Rb	85911170	None	18.7 days	β⁻ 1.774, 0.697; γ 1.077
	^{87}Rb	86.909187	27.8	Stable (?)	
Strontium, Z = 38	^{82}Sr	81.91840	None	25.4 days	EC
	^{83}Sr	82.91756	None	1.5 days	EC; β⁺ many; γ many
	^{84}Sr	83.913430	0.6	Stable	
	^{85}Sr	84.912936	None	64.9 days	EC; γ 0.514
	^{86}Sr	85.9092672	9.9	Stable	
	^{87}Sr	86.9088841	7.0	Stable	
	^{88}Sr	87.9056188	82.6	Stable	
	^{89}Sr	88.907455	None	50.5 days	β⁻ 1.488
	^{90}Sr	89.907738	None	29.1 years	β⁻ 0.546
Yttrium, Z = 39	^{88}Y	87.909506	None	106 days	EC; γ 1.863, 0.898
	^{89}Y	88.905849	100	Stable	
	^{90}Y	89.907301	None	2.67 days	β⁻ 2.280
Zirconium, Z = 40	^{90}Zr	89.9047026	51.5	Stable	
	^{91}Zr	90.9056439	11.2	Stable	
	^{92}Zr	91.9050386	17.2	Stable	
	^{93}Zr	92.906474	None	1.5×10^6 years	β⁻ 0.060; γ 0.031
	^{94}Zr	93.9063148	17.4	Stable	
	^{95}Zr	94.908041	None	64.0 days	β⁻ 0.400, 0.367; γ 0.757, 0.724
	^{96}Zr	95.908275	2.8	Stable	
	^{97}Zr	96.910950	None	16.8 h	β⁻ many; γ many
Niobium, Z = 41	^{92}Nb	91.907192	None	3.7×10^7 years	EC
	^{93}Nb	92.9063772	100	Stable	
	^{94}Nb	93.907282	None	2.4×10^4 years	β⁻ 0.471; γ 0.871, 0.702

(continued)

Element	Nuclide	Atomic Mass[a]	Abundance[b]	Half-Life	Radiations, Energies[c]
Molybdenum, Z = 42	^{92}Mo	91.906809	18.8	Stable	
	^{93}Mo	92.906811	None	3.5×10^3 years	EC; γ 0.031
	^{94}Mo	93.9050853	9.3	Stable	
	^{95}Mo	94.9058411	15.9	Stable	
	^{96}Mo	95.9046758	16.7	Stable	
	^{97}Mo	96.9060205	9.6	Stable	
	^{98}Mo	97.9054073	24.1	Stable	
	^{99}Mo	98.907711	None	2.7 days	β⁻ 1.214; γ 0.778, 0.141
	^{100}Mo	99.907477	9.6	Stable	
Technetium, Z = 43	^{96}Tc	95.90787	None	4.3 days	β⁺ many; γ many
	^{97}Tc	97.96906364	None	2.6×10^6 years	EC
	^{98}Tc	97.907215	None	4.2×10^6 years	β⁻ 0.397; γ 0.745, 0.652
	^{99}Tc	98.907	None	2.1×10^5 years	β⁻ 0.294
	^{99m}Tc	98.907	None	6.01 h	β⁻ 0.347
Ruthenium, Z = 44	^{96}Ru	96.95907599	5.5	Stable	
	^{97}Ru	96.90756	None	2.9 days	EC
	^{98}Ru	97.905287	1.9	Stable	
	^{99}Ru	98.90593889	12.8	Stable	
	^{100}Ru	99.9042192	12.8	Stable	
	^{101}Ru	100.9055819	17.1	Stable	
	^{102}Ru	101.9043485	31.6	Stable	
	^{103}Ru	102.906323	None	39.3 days	β⁻ 0.223; γ 0.497
	^{104}Ru	103.905425	18.6	Stable	
	^{105}Ru	104.907750	None	4.4 h	β⁻ 1.187; γ 0.469
Rhodium, Z = 45	^{102}Rh	101.908642	None	2.9 years	EC; γ 0.766, 0.697
	^{103}Rh	102.905500	100	Stable	
	^{104}Rh	103.906655	None	4.2 s	β⁻ 2.448
	^{105}Rh	104.905692	None	35.4 h	β⁻ 0.566, 0.248; γ 0.319, 0.306

Element	Nuclide	Atomic Mass[a]	Abundance[b]	Half-Life	Radiations, Energies[c]
Palladium, Z = 46	[102]Pd	101.905634	1.0	Stable	
	[103]Pd	102.906087	None	17.0 days	EC; γ 0.040
	[104]Pd	103.904029	11.1	Stable	
	[105]Pd	104.905079	22.3	Stable	
	[106]Pd	105.903478	27.3	Stable	
	[107]Pd	106.90513	None	6.5×10^6 years	β⁻ 0.033
	[108]Pd	107.3895	26.5	Stable	
	[109]Pd	108.905954	None	13.5 h	β⁻ 1.028
	[110]Pd	109.905167	11.7	Stable	
Silver, Z = 47	[107]Ag	106.905092	51.8	Stable	
	[108]Ag	107.5654	None	2.4 min	β⁻ 1.650
	[109]Ag	108.904756	48.2	Stable	
	[110]Ag	109.906111	None	25 s	β⁻ 2.891
Cadmium, Z = 48	[106]Cd	105.906461	1.3	Stable	
	[107]Cd	106.90661	None	6.5 h	EC; γ 0.093
	[108]Cd	107.904176	0.9	Stable	
	[109]Cd	108.90495	None	462 days	EC
	[110]Cd	109.90305	12.5	Stable	
	[111]Cd	110.904182	12.8	Stable	
	[112]Cd	111.902757	24.1	Stable	
	[113]Cd	112.904400	12.2	Stable	
	[114]Cd	113.903337	28.7	Stable	
	[115]Cd	114.905431	None	22 h	β⁻ 1.110, 0.593; γ 0.527
	[116]Cd	115.904755	7.5	Stable	
Indium, Z = 49	[112]In	111.90553	None	14.4 min	EC; β⁺ 2.580; γ 0.606
	[113]In	112.904061	4.3	Stable	
	[114]In	113.904918	None	1.2 min	β⁻ 1.984
	[115]In	114.903882	95.7	4.4×10^{14} years	β⁻ 0.482
	[116]In	115.905260	None	54.3 min	β⁻ 1.010, 0.872; γ 1.097, 0.417
Tin, Z = 50	[112]Sn	111.904826	1.0	Stable	
	[113]Sn	112.905174	None	115 days	EC; γ 0.931
	[114]Sn	113.902748	0.7	Stable	
	[115]Sn	114.903348	0.3	Stable	
	[116]Sn	115.901747	14.5	Stable	

(continued)

Element	Nuclide	Atomic Mass[a]	Abundance[b]	Half-Life	Radiations, Energies[c]
	117Sn	116.902956	7.7	Stable	
	118Sn	117.901609	24.2	Stable	
	119Sn	118.903311	8.6	Stable	
	120Sn	119.9021191	32.6	Stable	
	121Sn	120.904239	None	1.1 days	β⁻ 0.389
	122Sn	121.9034404	4.6	Stable	
	123Sn	122.905723	None	129 days	β⁻ 1.403
	124Sn	123.9052743	5.8	Stable	
Antimony, Z = 51	121Sb	120.9038212	57.2	Stable	
	122Sb	121.990518	None	2.7 days	β⁻ 1.980, 1.414; γ 0.564
	123Sb	122.9042160	42.8	Stable	
	124Sb	123.905938	None	60.3 days	β⁻ 2.301, 0.608; γ 0.603
Tellurium, Z = 52	120Te	119.904048	0.1	Stable	
	121Te	120.90494	None	16.8 days	EC; γ 0.573
	122Te	121.903050	2.6	Stable	
	123Te	122.9042719	0.9	Stable	
	124Te	123.9028180	4.7	Stable	
	125Te	124.9044285	7.1	Stable	
	126Te	125.9033095	18.8	Stable	
	127Te	126.905217	None	9.4 h	β⁻ 0.698
	128Te	127.904463	31.7	Stable	
	129Te	128.906596	None	33.6 days	β⁻ many; γ many
	130Te	129.906229	30.1	Stable	
Iodine, Z = 53	125I	124.904624	None	59.4 days	EC
	126I	125.905613	None	13.0 days	EC; γ 0.666
	127I	126.904473	100	Stable	
	128I	127.905805	None	25.0 min	β⁻ 2.119; γ 0.442
	129I	128.904988	None	1.6×10^7 years	β⁻ 0.154
	130I	129.906674	None	12.4 h	β⁻ many; γ many
	131I	130.906125	None	8.0 days	β⁻ many; γ many
Xenon, Z = 54	124Xe	123.905842	0.1	Stable	
	125Xe	124.906398	None	17.1 h	EC
	126Xe	125.904281	0.1	Stable	
	127Xe	126.905179	None	36.4 days	EC; γ 0.202

Element	Nuclide	Atomic Mass[a]	Abundance[b]	Half-Life	Radiations, Energies[c]
	^{128}Xe	127.935312	1.9	Stable	
	^{129}Xe	128.9047801	26.4	Stable	
	^{130}Xe	129.9035094	4.1	Stable	
	^{131}Xe	130.905072	21.2	Stable	
	^{132}Xe	131.904144	26.9	Stable	
	^{133}Xe	132,905906	None	5.2 days	β^- 0.154
	^{134}Xe	133.905395	10.4	Stable	
	^{135}Xe	134.90721	None	9.1 h	β^- 0.910
	^{136}Xe	135.907214	8.9	Stable	
Cesium, Z = 55	^{132}Cs	131.906430	None	6.5 days	EC; γ 0.667
	^{133}Cs	132.905429	100	Stable	
	^{134}Cs	133.906714	None	8.5 days	β^- 0.658; γ 0.604
	^{135}Cs	134.905972	None	2.3×10^6 years	β^- 0.269
	^{136}Cs	135.907307	None	13.2 days	β^- 0.341; γ 0.818
	^{137}Cs	136.907085	None	30.2 years	β^- 0.514; γ 0.662
Barium, Z = 56	^{130}Ba	129.906282	0.1	Stable	
	^{131}Ba	130.90693	None	11.7 days	EC; γ many
	^{132}Ba	131.905042	0.1	Stable	
	^{133}Ba	132.906003	None	10.5 years	EC; γ 0.356
	^{134}Ba	133.904486	2.4	Stable	
	^{135}Ba	134.905665	6.6	Stable	
	^{136}Ba	135.904553	7.9	Stable	
	^{137}Ba	136.905812	11.2	Stable	
	^{138}Ba	137.905232	71.7	Stable	
	^{139}Ba	138.908836	None	1.4 h	β^- 2.316, 2.141; γ 0.166
	^{140}Ba	139.91060	None	12.8 days	β^- 1.020, 1.003, 0.480; γ 0.537, 0.030
Lanthum, Z = 57	^{138}La	137.907105	0.1	Stable	
	^{139}La	138.906342	99.9	Stable	
	^{140}La	139.909473	None	1.7 days	β^- many; γ many
Cerium, Z = 58	^{136}Ce	135.907140	0.2	Stable	
	^{137}Ce	136.90788	None	9.0 h	EC; γ many
	^{138}Ce	137.905985	0.3	Stable	

(continued)

Element	Nuclide	Atomic Mass[a]	Abundance[b]	Half-Life	Radiations, Energies[c]
	^{139}Ce	138.90665	None	138 days	EC; γ 0.166
	^{140}Ce	139.905433	88.5	Stable	
	^{141}Ce	140.908272	None	32.5 days	β⁻ 0.437; γ 0.145
	^{142}Ce	141.909241	11.1	Stable	
	^{143}Ce	142.912382	None	1.4 days	β⁻ 1.110; γ 0.293
Praseodymium, Z = 59	^{141}Pr	140.907647	100	Stable	
	^{142}Pr	141.910041	None	19.1 h	β⁻ 2.158, 0.580; γ 1.575
	^{143}Pr	142.910813	None	13.6 days	β⁻ 0.934
Neodymium, Z = 60	^{142}Nd	141.907719	27.2	Stable	
	^{143}Nd	142.909810	12.2	Stable	
	^{144}Nd	143.90083	23.8	Stable	
	^{145}Nd	144.912570	8.3	Stable	
	^{146}Nd	145.913113	17.2	Stable	
	^{147}Nd	146.916096	None	11.0 days	β⁻ 0.804; γ 0.531
	^{148}Nd	147.916889	5.7	Stable	
	^{149}Nd	148.920145	None	1.7 h	β⁻ many; γ many
	^{150}Nd	149.920887	5.6	Stable	
Promethium, Z = 61	^{145}Pm	144.912743	None	17.7 years	EC; γ 0.072
	^{146}Pm	145.914693	None	5.5 years	EC; γ 0.454
	^{147}Pm	146.915134	None	2.6 years	β⁻ 0.225
	^{148}Pm	147.91747	None	5.4 days	β⁻ 0.416; γ 0.556
	^{149}Pm	148.918330	None	2.2 days	β⁻ 1.067; γ 0.286
	^{150}Pm	149.92098	None	2.7 h	β⁻ many; γ many
	^{151}Pm	150.92120	None	1.2 days	β⁻ many; γ many
Samarium, Z = 62	^{144}Sm	143.911998	3.1	Stable	
	^{145}Sm	144,913407	None	340 days	EC; γ 0.061
	^{146}Sm	145.913038	None	1.0×10^8 years	γ 2.46
	^{147}Sm	146.914894	15.0	Stable	
	^{148}Sm	147.914819	11.2	Stable	
	^{149}Sm	148.917180	13.8	Stable	

Element	Nuclide	Atomic Mass[a]	Abundance[b]	Half-Life	Radiations, Energies[c]
	^{150}Sm	149.917273	7.4	Stable	
	^{151}Sm	150.919929	None	90 years	β^- 0.075
	^{152}Sm	151.919728	26.8	Stable	
	^{153}Sm	152.992094	None	1.9 days	β^- many; γ many
	^{154}Sm	153.922205	22.8	Stable	
Europium, Z = 63	^{151}Eu	150.919702	47.8	Stable	
	^{152}Eu	151.921741	None	15.5 years	EC; γ 0.122
	^{152m}Eu	151.921741	None	9.3 h	β^- 1.864
	^{153}Eu	152.921225	52.2	Stable	
Gadolinium, Z = 64	^{152}Gd	151.919786	0.2	Stable	
	^{153}Gd	152.921747	None	242 days	EC; γ 0.103
	^{154}Gd	153.920861	2.2	Stable	
	^{155}Gd	154.922618	14.8	Stable	
	^{156}Gd	155.922118	20.5	Stable	
	^{157}Gd	156.923956	15.7	Stable	
	^{158}Gd	157.924019	24.8	Stable	
	^{159}Gd	158.926385	None	18.6 h	β^- 0.960; γ 0.364
	^{160}Gd	159.927049	21.7	Stable	
Terbium, Z = 65	^{158}Tb	157.925410	None	110 years	EC; γ 0.944
	^{159}Tb	158.925342	100	Stable	
	^{160}Tb	159.927164	None	72.3 days	β^- 0.869, 571; γ 0.966, 0.879
Dysprosium, Z = 66	^{156}Dy	155.924277	0.1	Stable	
	^{157}Dy	156.92546	None	8.1 h	EC; γ 0.326
	^{158}Dy	157.924403	0.1	Stable	
	^{159}Dy	158.927536	None	144 days	EC
	^{160}Dy	159.925193	2.3	Stable	
	^{161}Dy	160.926930	18.9	Stable	
	^{162}Dy	161.926795	25.5	Stable	
	^{163}Dy	162.928728	24.9	Stable	
	^{164}Dy	163.929171	28.2	Stable	
	^{165}Dy	164.927100	None	2.3 h	β^- 1.285; γ 0.095
Holmium, Z = 67	^{163}Ho	162.928730	None	4.6×10^3 years	EC
	^{164}Ho	163.930231	None	29 min	EC; γ 0.074
	^{165}Ho	164.930319	100	Stable	

(continued)

Element	Nuclide	Atomic Mass[a]	Abundance[b]	Half-Life	Radiations, Energies[c]
	^{166}Ho	165.932281	None	1.1 days	β⁻ 1.854, 1.773
Erbium, Z = 68	^{162}Er	161.928775	0.1	Stable	
	^{163}Er	162.93003	None	1.3 h	EC
	^{164}Er	163.929198	1.6	Stable	
	^{165}Er	164.930723	None	10.4 h	EC
	^{166}Er	165.930290	33.6	Stable	
	^{167}Er	166.932046	22.9	Stable	
	^{168}Er	167.932368	26.8	Stable	
	^{169}Er	186.934588	None	9.4 days	β⁻ 0.342, 0.350
	^{170}Er	169.934561	14.9	Stable	
	^{171}Er	170.938026	None	7.5 h	β⁻ 1.065; γ 0.308
	^{172}Er	171.939352	None	2.1 days	β⁻ 0.381, 0.279; γ 0.610, 0.407
Thulium, Z = 69	^{168}Tm	167.934271	None	93.1 days	EC; γ 0.731
	^{169}Tm	168.934212	100	Stable	
	^{170}Tm	169.935798	None	129 days	β⁻ 0.968 0.883; γ 0.084
Ytterbium, Z = 70	^{168}Yb	167.933804	0.1	Stable	
	^{169}Yb	168.935187	None	32.0 days	EC; γ 0.063
	^{170}Yb	169.934759	3.0	Stable	
	^{171}Yb	170.936323	14.3	Stable	
	^{172}Yb	171.936378	21.8	Stable	
	^{173}Yb	172.938208	16.1	Stable	
	^{174}Yb	173.938859	31.8	Stable	
	^{175}Yb	174.941273	None	4.2 days	β⁻ 0.468; γ 0.396
	^{176}Yb	175.942564	12.8	Stable	
	^{177}Yb	176.945257	None	1.9 h	β⁻ 1.400, 1.129; γ 1.080, 0.150
Lutetium, Z = 71	^{175}Lu	174.940770	97.4	Stable	
	^{176}Lu	175.942679	2.6	Stable	
	^{177}Lu	176.943755	None	6.7 days	β⁻ 0.498, 0.385, 0.176; γ 0.208
Hafnium, Z = 72	^{172}Hf	171.93946	None	1.9 years	EC; γ 0.023
	^{173}Hf	172.9407	None	23.6 h	EC; γ 0.124

Element	Nuclide	Atomic Mass[a]	Abundance[b]	Half-Life	Radiations, Energies[c]
	174Hf	173.940044	0.2	Stable	
	175Hf	174.941504	None	70 days	EC; γ 0.343
	176Hf	175.941406	5.3	Stable	
	177Hf	176.943217	18.6	Stable	
	178Hf	177.943696	27.3	Stable	
	179Hf	178.9458122	13.6	Stable	
	180Hf	179.9456457	35.1	Stable	
	181Hf	180.949099	None	42.4 days	β^- 0.405; γ 0.482
	182Hf	181.95055	None	9×10^6 years	β^- 0.103; γ 0.270
Tantalum, Z = 73	179Ta	178.94593	None	1.8 years	EC
	180Ta	179.947462	Trace	Stable	
	181Ta	180.947992	100	Stable	
	182Ta	181.950152	None	114 days	β^- 0.531; γ 1.121, 0.068
Tungsten, Z = 74	180W	179.946701	0.1	Stable	
	181W	180.94820	None	121 days	EC
	182W	181.948202	26.5	Stable	
	183W	182.950220	14.3	Stable	
	184W	183.959028	30.6	Stable	
	185W	184.953420	None	74.8 days	β^- 0.432
	186W	185.954357	28.4	Stable	
	187W	186.957158	None	23.9 h	β 0.622; γ 0.134, 0.072
	187W	187.958487	None	69.4 days	β^- 0.349
Rhenium, Z = 75	184Re	183.95252	None	70 days	EC
	185Re	184.952951	37.4	Stable	
	186Re	185.954986	None	3.8 days	β^- 0.1.071, 0.934
	187Re	186.955744	62.6	Stable	
	188Re	187.958112	None	16.9 h	β^- 2.120; γ 0.632, 0.478
Osmium, Z = 76	184Os	183.952488	Trace	Stable	
	185Os	184.954043	None	93.6 days	EC
	186Os	185.953830	1.6	Stable	
	187Os	186.955741	2.0	Stable	
	188Os	187.955830	13.2	Stable	
	189Os	188.958137	16.2	Stable	

(continued)

Element	Nuclide	Atomic Mass[a]	Abundance[b]	Half-Life	Radiations, Energies[c]
	^{190}Os	189.958436	26.3	Stable	
	^{191}Os	190.960928	None	15.4 days	β⁻ 0.143; γ 0.129
	^{192}Os	191.961467	40.8	Stable	
	^{193}Os	192.964138	None	30.5 h	β⁻ 1.140; γ 0.460, 0.139
Iridium, Z = 77	^{191}Ir	190.960584	37.3	Stable	
	^{192}Ir	191.962603	None	73.8 days	β⁻ 0.672, 0.535; γ 0.486, 0.316
	^{193}Ir	192.962917	62.7	Stable	
Platinum, Z = 78	^{190}Pt	189.959917	Trace	Stable	
	^{191}Pt	190.961684	None	3.0 days	EC
	^{192}Pt	191.961019	0.8	Stable	
	^{193}Pt	192.962984	None	50 years	EC
	^{194}Pt	193.962655	33.0	Stable	
	^{195}Pt	194.964766	34.0	Stable	
	^{196}Pt	195.964926	25.2	Stable	
	^{197}Pt	196.967232	None	18.9 h	β⁻ 0.642; γ 0.191
	^{198}Pt	197.967869	7.1	Stable	
Gold, Z = 79	^{195}Au	194.965017	None	186 days	EC; γ 0.099
	^{196}Au	195.966551	None	6.2 days	EC; γ 0.355
	^{197}Au	196.966543	100	Stable	
	^{198}Au	197.968225	None	2.7 days	β⁻ 0.962; γ 0.412
Mercury, Z = 80	^{196}Hg	195.965807	0.1	Stable	
	^{197}Hg	196.967195	None	2.7 days	EC; γ 0.077
	^{198}Hg	197.966743	10.0	Stable	
	^{199}Hg	198.968254	16.9	Stable	
	^{200}Hg	199.968300	23.1	Stable	
	^{201}Hg	200.970277	13.2	Stable	
	^{202}Hg	201.970617	29.9	Stable	
	^{203}Hg	202.972857	None	46.6 days	β⁻ 0.492; γ 0.279
	^{204}Hg	2.3.973467	6.9	Stable	
Thallium, Z = 81	^{203}Tl	202.972320	29.5	Stable	
	^{204}Tl	203.97385	None	3.8 years	β⁻ 0.763
	^{205}Tl	204.974401	70.5	Stable	
	^{206}Tl	205.976095	None	4.2 min	β⁻ 1.527
Lead, Z = 82	^{204}Pb	203.973020	1.4	Stable	

Element	Nuclide	Atomic Mass[a]	Abundance[b]	Half-Life	Radiations, Energies[c]
	^{205}Pb	204.97447	None	1.5×10^7 years	EC
	^{206}Pb	205.974440	24.1	Stable	
	^{207}Pb	206.975872	22.4	Stable	
	^{208}Pb	207.976627	52.4	Stable	
	^{209}Pb	208.981075	None	3.3 h	β⁻ 0.644
	^{210}Pb	209.98417	Trace	22.6 years	β⁻ 0.016
	^{211}Pb	210.98873	None	36.1 min	β⁻ 1.378
	^{212}Pb	211.99187	None	10.6 h	β⁻ 0.335; γ 0.239
	^{213}Pb	212.99650	None	10.2 min	β⁻ 1.980
	^{214}Pb	213.99980	None	26.8 min	β⁻ 0.728 0.670; γ 0.351, 0.295
Bismuth, Z = 83	^{208}Bi	207.97973	None	3.7×10^5 years	EC; γ 2.610
	^{209}Bi	208.980374	100	Stable	
	^{210}Bi	209.98410	None	5.1 days	β⁻ 1.161
Polonium, Z = 84	^{208}Po	207.90123	None	2.9 years	α 5.115
	^{209}Po	208.982404	None	102 years	α 4.884
	^{210}Po	209.98286	None	134 days	α 5.304
Astatine, Z = 85	^{209}At	208.98616	None	5.4 h	EC
	^{210}At	209.987126	None	8.1 h	EC; γ 1.181
	^{211}At	210.98748	None	7.2 h	EC; γ 0.687
Radon, Z = 86	^{221}Rn	221.0156	None	25 min	β⁻ 0.855; γ 0.186
	^{222}Rn	222.017571	None	3.8 days	α 5.489
	^{223}Rn	223.0218	None	23.2 min	β⁻ 0.592
	^{224}Rn	224.0241	None	107 min	β⁻ 0.265 0.260
Francium, Z = 87	^{222}Fr	222.01754	None	14.3 min	β⁻ many; γ many
	^{223}Fr	223.01973	None	22.0 min	β⁻ 1.097; γ 0.050
	^{224}Fr	224.02323	None	3.3 min	β⁻ 2.614; γ 0.131
Radium, Z = 88	^{223}Ra	223.018497	None	11.4 days	α 5.716, 5.607
	^{224}Ra	224.020202	None	3.7 days	α 5.685
	^{225}Ra	225.923603	None	14.9 days	β⁻ 0.771, 0.320; γ 0.040
	^{226}Ra	226.025403	100	1,599 years	α 4.784

(continued)

Element	Nuclide	Atomic Mass[a]	Abundance[b]	Half-Life	Radiations, Energies[c]
	^{227}Ra	227.029170	None	42 min	β⁻ 1.296, 1.277; γ 0.027
	^{228}Ra	228.031063	None	5.8 years	β⁻ 0.039, 0.025; γ 0.013
Actinium, Z = 89	^{225}Ac	225.02322	None	10.0 days	α 5.830
	^{226}Ac	226.026089	None	1.2 days	β⁻ 1.105, 0.885; γ 0.230
	^{227}Ac	227.027750	None	21.8 years	β⁻ 0.044, 0.035, 0.020; γ 0.025, 0.015, 0.009
	^{228}Ac	228.031104	None	6.2 h	β⁻ many; γ many
Thorium, Z = 90	^{229}Th	229.031754	None	7.9×10^3 years	α 4.845; γ many
	^{230}Th	230.033126	None	7.5×10^4 years	α 4.687, 4.620; γ 0.067
	^{231}Th	231.036296	None	1.1 days	β⁻ 0.305, 0.288; γ 0.026
	^{232}Th	232.0380508	100	1.4×10^{10} years	α 4.012, 3.974
	^{233}Th	233.041576	None	22.3 min	β⁻ 1.243, 1.236; γ many
	^{234}Th	234.036596	None	21.1 days	β⁻ 0.199, 0.104; γ 0.093
Protactinium, Z = 91	^{230}Pa	230.03453	None	17.4 days	EC; γ 0.951, 0.918
	^{231}Pa	231.035880	None	3.3×10^4 years	α 5.028, 5.014, 4.951; Γ 0.283, 0.145, 0.129
	^{232}Pa	232.0358	None	1.3 days	β⁻ 0.314, 0.294; γ 0.894
	^{233}Pa	233.04024	None	27.0 days	β⁻ 0.231, 0.156; γ 0.312
Uranium, Z = 92	^{234}U	234.0404946	Trace	2.5×10^5 years	α 4.774
	^{235}U	235.0439231	0.7	7.0×10^8 years	α 4.397

Element	Nuclide	Atomic Mass[a]	Abundance[b]	Half-Life	Radiations, Energies[c]
	^{236}U	236.045562	None	2.3×10^7 years	α 4.494
	^{237}U	237.0487240	None	6.8 days	β⁻ 0.251, 0.237; γ 0.208, 0.060
	^{238}U	238.0507826	99.3	4.5×10^9 years	α 4.189, 4.151; γ 0.049
Neptunium, Z = 93	^{237}Np	237.04817	None	2.1×10^6 years	α 4.788, 4.771
	^{238}Np	238.05094	None	2.1 days	β⁻ 0.262
	^{239}Np	239.502931	None	2.4 days	β⁻ 0.438, 0.341; γ 0.106
Plutonium, Z = 94	^{238}Pu	238.04955	None	87.7 years	α 5.593
	^{239}Pu	239.05216	None	2.4×10^4 years	α 5.156
	^{242}Pu	242.05874	None	3.8×10^5 years	α 4.900
	^{244}Pu	244.0641	None	8.2×10^7 years	α 4.589
Americium, Z = 95	^{240}Am	240.05529	None	2.1 days	EC
	^{241}Am	241.05682	None	432 years	α 5.485
	^{242}Am	242.05654	None	16.0 h	β⁻ 0.662, 0.320; γ 0.042
	^{243}Am	243.061375	None	7.37×10^3 years	α 5.275
Curium, Z = 96	^{241}Cm	241.05765	None	32.8 days	EC
	^{243}Cm	243.06138	None	28.5 years	α 5.785
	^{245}Cm	245.06548	None	8.50×10^3 years	α 5.362
	^{247}Cm	247.070347	None	1.56×10^7 years	α 4.870
Berkelium, Z = 97	^{246}Bk	246.0687	None	1.8 days	EC
	^{247}Bk	247.07030	None	1.4×10^3 years	α 5.531, 5.710; γ 0.265
	^{248}Bk	248.07310	None	23.7 h	α 5.803
	^{249}Bk	249.07498	None	320 days	β⁻ 0.124
Californium, Z = 98	^{251}Cf	251.079580	None	898 years	α 5.854, 5.8679; γ 0.176

(continued)

Element	Nuclide	Atomic Mass[a]	Abundance[b]	Half-Life	Radiations, Energies[c]
	^{252}Cf	252.08162	None	2.64 years	α 6.118; SF 3%
	^{253}Cf	253.08513	None	17.8 days	β⁻ 0.270
Einsteinium, Z = 99	^{249}Es	249.07640	None	1.7 days	EC; γ 0.379
	^{252}Es	252.082944	None	1.29 years	α 6.632
	^{255}Es	255.09027	None	40 days	β⁻ 0.300
Fermium, Z = 100	^{251}Fm	251.08157	None	5.3 h	EC
	^{253}Fm	253.08517	None	3 days	EC; γ 0.050
	^{257}Fm	257.095099	None	101 days	α 6.520
Mendelevium, Z = 101	^{255}Md	255.09108	None	27 min	EC; γ 0.282
	^{258}Md	258.09857	None	51.5 days	EC
	^{259}Md	259.1005	None	1.6 h	SF
Nobelium, Z = 102	^{253}No	253.0907	None	1.7 min	α 8.010
	^{256}No	256.0943	None	2.9 s	α 8.448
	^{259}No	259.100931	None	58 min	α 7.500
Lawrencium, Z = 103	^{255}Lr	255.0967	None	22 s	α 8.410, 8.360
	^{262}Lr	262.1097	None	3.6 h	EC
Rutherfordium, Z = 104	^{259}Rf	259.1056	None	3.4 s	α
	^{261}Rf	261.10869	None	1.1 min	α
	^{263}Rf	263.1125	None	10 min	α
Dubnium, Z = 105	^{255}Db	255.1074	None	1.6 s	α
	^{263}Db	263.1153	None	30 s	α; SF
Seaborgium, Z = 106	^{258}Sg	258.11317	None	3.3 ms	SF
	^{259}Sg	259.11450	None	580 ms	α
	^{569}Sg	269.12876	None	35 s	α
Bohrium, Z = 107	^{260}Bh	260.12197	None	300 μs	α
	^{266}Bh	266.12694	None	5 s	α
	^{273}Bh	273.13962	None	90 min	α
Hassium, Z = 108	^{265}Hs	265.13009	None	780 μs	α
	^{270}Hs	270.13465	None	30 s	α
	^{274}Hs	274.14313	None	1 h	α
Meitnerium, Z = 109	^{265}Mt	265.13615	None	2 ms	α
	^{274}Mt	274.14749	None	20s	α
	^{278}Mt	278.15481	None	30 min	α

Element	Nuclide	Atomic Mass[a]	Abundance[b]	Half-Life	Radiations, Energies[c]
Darmstadtium, Z = 110	^{267}Ds	267.14434	None	10 μs	α
	^{274}Ds	274.14949	None	2 s	α
	^{281}Ds	281.16206	None	4 min	α
Roentgenium, Z = 111	^{272}Rg	272.15362	None	2 ms	α
	^{277}Rg	277.15952	None	1 s	α
	^{283}Rg	283.16842	None	10 min	α
Copernicium, Z = 112	^{277}Cn	277.16394	None	1.1 ms	α
	^{279}Cn	279.16655	None	100 ms	α
Ununtrium, Z = 113	^{283}Uut	283.17645	None	150 ms	α
	^{284}Uut	284.17808	None	~1 s	α
Ununquadium, Z = 114	^{288}Uuq	288.18569	None	~5 s	α
	^{289}Uuq	289.18728	None	21 s	α
Ununpentium, Z = 115	^{287}Uup	287.18560	None	~50 ms	α
	^{288}Uup	288.18569	None	~2 ms	α
Ununhexium, Z = 116	^{290}Uuh	290.19859	None	~30 ms	α
	^{292}Uuh	292.19979	None	~50 ms	α
Ununseptium, Z = 117	^{291}Uus	291.20656	None	10 ms	α
	^{292}Uus	292.2075	None	50 ms	α
Ununoctium, Z = 118	^{294}Uuo	294.2147	None	~1 ms	α

[a] The atomic mass of a nuclide includes the mass of Z electrons.
[b] The abundance of a nuclide is its mass percent present in nature.
[c] The radiations listed for a nuclide are those most frequently observed in its radioactive decay.

Appendix 3: some notes

Experiment 1

The halogen-quenched, end-window Geiger-Müller detector typically has an operating potential of 900 to 1,000 volts. The efficiency of the Geiger-Müller detector is calculated from 100(cpm/dpm), where cpm is the observed count rate and dpm is the actual activity of the radioactive source calculated from the number of becquerel (Bq) or microcurie (µCi). One Bq is one dps, and one µCi is 3.7×10^4 dps. Typically, efficiency is approximately 5% for a hard ($E_{max} \geq 1$ MeV) beta emitter 2 cm from the detector. For sealed sources covered with approximately 1 mg/cm² of Mylar™ counted 2 cm from a Geiger-Müller detector having a 1.4 mg/cm² window, the counting efficiencies for ^{12}C ($E_{max} = 0.156$ MeV), ^{210}Bi ($E_{max} = 1.16$ MeV from RaDEF), and ^{90}Y ($E_{max} = 2.27$ MeV from ^{90}Sr) were 0.8, 4.2, and 8.2%, respectively.

Experiment 2

The split-source kit (such as RSS-2 from Spectrum Technologies) consisting of two semicircular sources, S_1 and S_2, and a semicircular blank, B, is suitable for this experiment. Using the equations given in the narrative to Experiment 2, resolving times for halogen-quenched Geiger-Müller detectors typically range from 200 to 450 µs.

Experiment 3

The typical background for a halogen-quenched, end-window Geiger-Müller detector is approximately 30 cpm. This can be reduced to approximately 15 cpm with 1 inch of lead shielding.

Experiment 4

This experiment is conveniently conducted using a meter stick to measure the distance between the Geiger-Müller detector and the radioactive material. A 10-μCi ^{90}Sr sealed source is recommended. It is important to maintain a rigid alignment of the detector with the meter stick during this experiment.

Deviations from linearity may occur at the ends of the curve. If observed, they may be attributed to statistical variations at low count rates and dead time losses at high count rates.

Experiment 5

The determination of shelf ratios is another example of the inverse square law described in Experiment 4. However, other factors, such the scatter and the energy of the beta radiation, influence the results. In addition to showing the possibilities for increasing or decreasing count rates, this experiment clearly demonstrates the need to maintain constant counting geometry in order to obtain reproducible results.

Typically, this experiment will show that count rates from low-energy beta emitters such as ^{14}C are more affected by movement from one shelf to another than are the count rates from high-energy beta emitters such as ^{32}P and ^{90}Sr.

Experiment 6

This experiment can be carried out using the S-10 back scatter kit from Canberra Industries. Typical back scatters of radiation from ^{32}P are 1.25 increase with 2-mm aluminum, 1.35 increase with 2-mm iron, 1.45 increase with 2-mm copper, and 1.75 increase with 2-mm lead.

Experiment 7

As with all experiments involving radioactivity measurements on liquid samples, a spill and the consequent contamination of the laboratory and the equipment is a possibility. A practice run with distillation of the pipetting, transfer, and counting is recommended.

Contamination of the laboratory and the equipment with uranyl nitrate can be monitored using its characteristic yellow-green fluorescence under ultraviolet illumination.

In Part A of this experiment, both the total activity and the total volume of the samples increase. This is accompanied by a nonlinear increase of the activity.

In Part B of this experiment, the total activity of the sample remains constant while its volume increases. This is accompanied by a nonlinear decrease in the activity.

In both cases, the nonlinearity can be attributed to self-absorption.

Experiment 8

The method of Feather begins with plotting the logarithm of the corrected activity against absorber thickness in mg/cm² and extrapolating the end of the range of the beta radiation from the RaDEF reference to 476 mg/cm². The abscissa of this attenuation curve is then divided into 10 tenths of 47.6 mg/cm² each. A transmission factor for each tenth of the range is calculated from the ratio of the activity at that tenth to the activity at zero absorber thickness, remembering to divide the activities, not their logarithms. The activity of the other beta emitter at zero absorber thickness is multiplied by each of the transmission factors as shown below:

Total absorber thickness, mg/cm²	Corrected RaDEF activity, cpm	Corrected sample activity, cpm
4.0	62,045	27,847
25.4	4,0514	24,366
45.3	28,501	21,891
73.8	22,670	20,305
102.5	11,987	15,925
143.2	5,797	11,904
180.6	2,305	8,061
229.7	1,688	6,833
273.1	765	4,937
315.9	245	3,045
373.7	97	2,002
444.5	23	981
540.6	22	212

Tenth of Range	RaDEF Activity	Transmission Factor	Sample X Factor R_0	Absorber Thickness	Apparent Range
0	62,000	1.00	28,000		
1	26,600	0.429	12,012	136	1,360
2	12,300	0.198	5,544	252	1,260
3	5,000	0.0806	2,257	361	1,203
4	2,300	0.0370	1,036	437	1,093
5	1,040	0.0167	468	551	1,102
6	400	0.00645	181	609	1,015
7	136	0.00219	61		
8					

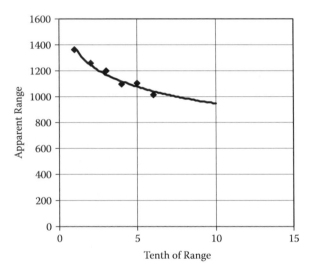

Figure 32 Beta range extrapolation.

Extrapolation of the apparent range to the 10th tenth indicates a range of 950 mg/cm². The results shown in the Figure 32 were obtained with [90]Sr. Values ranging from 650 to 800 mg/cm² are obtained with [32]P.

Experiment 9

All Foils, Inc. is among the suppliers of thin metal foils. For this experiment, foils of a few tenths of a millimeter thickness are recommended. The thickness of the foils can be assumed to conform to the vendor's specifications, or it can be measured as a part of the experiment. The E_{max} of the emitters should be between 0.5 and 1.5 meV.

The activity decreases exponentially with increasing foil thickness up to a thickness sufficient to completely absorb the beta radiation. Foils of material having higher atomic numbers are more efficient in absorbing beta radiations. The attenuation of radiation initially of intensity I_0 as it transverses matter is exponentially dependent upon both the density of the medium and the thickness of the medium: $I = I_0 e^{-\mu \rho x}$, where μ is the absorption or attenuation coefficient, ρ is the density, and x is the thickness.

Experiment 10

The attenuation of gamma radiation is exponentially dependent upon the energy of the radiation, the density of the medium, and the thickness of the medium: $I = I_0 e^{-\mu \rho x}$, where μ is the absorption or attenuation coefficient, ρ is the density, and x is the thickness. The linear attenuation coefficient is inversely proportional to the half value layer (HVL); $\mu = 0.693/\text{HVL}$. Some half value layers calculated from the NISH (1996) tables are as follows:

Gamma Energy	Aluminum Absorber	Lead Absorber
0.1 MeV	1.3 cm	0.01 cm
0.5 MeV	3.1 cm	0.38 cm
1.0 Mev	4.2 cm	0.86 cm
1.5 Mev	5.1 cm	1.18 cm

Typical experimental HVLs in lead for ^{133}Ba, ^{137}Cs, and ^{60}Co range from 0.17 to 0.25, 0.60 to 0.74, and 1.2 to 1.6 cm, respectively.

Experiment 12

Generator kits for this experiment are available from Spectrum Techniques (part number: ISO) and other vendors.

The filter paper is used to break the surface tension of the water. It spreads the sample out, minimizing self-absorption.

G-M detectors can be used for this experiment, despite their low efficiency in counting gamma photons.

Experiment 13

Counting the rapidly decaying silver ^{110}Ag can be accomplished with a team of three students. One must watch the clock and announce the 10-second intervals. Another must watch the scaler and announce the total accumulated counts when the 10-second intervals are announced. The third must record the data.

Experiment 14

The mass of KCl should be determined using a balance capable of weighing to the nearest milligram. The mass must be accurately determined, but anywhere from 280 to 320 mg of KCl will work.

Dissolving the KCl in the planchette then drying under a heat lamp does not work well. Spreading the sample out in the planchette is the best way to reduce self-absorption.

Samples should be counted as close to the detector as possible. This is because count rates will be low, and minimizing distance to the detector will maximize counting efficiency.

^{210}Bi (or RaDEF) works well to determine efficiency.

Branched decay only affects half-life determinations like this one, where activity is being measured. When half-life is determined by measuring the change in count rate over time, branching in the decay is irrelevant. An analogy for this experiment is trying to determine when a bucket of water will be emptied by measuring the flow rate out of just one of two holes in the bucket. The time required will always be overestimated.

Possible systematic errors in this experiment are self-absorption, spilling, the percent efficiency value used, and the fact that the sample is a non-point source. All of these could lower the observed activity. Percent efficiency is the only one that could increase observed activity.

Experiment 15

Kodak BioMax XAR-2 general purpose autoradiographic film 1660760 is among those suitable for this experiment. Double-sided, paper-wrapped, 5 × 7 inch sheets are available from Sigma-Aldrich as well as from other venders. Sigma-Aldrich is a supplier of Kodak developer 1900943 and Kodak fixer 1901875 too.

Experiment 16

The Bendix Model 1050 Radioassay Electroscope was used in this experiment.

The results of this experiment clearly demonstrate the differences in specific ionization by α, β, and γ radiation. The respective discharge rates of the Bendix electroscope were approximately 10,000, 300, and 1 div/min/μCi with ^{241}Am, ^{204}Tl, and ^{133}Ba. This experiment can be elaborated upon to show a linear response using sealed 0.5 and 1 μCi ^{204}Tl sources.

Experiment 17

The results of this experiment show much lower count rates at the alpha plateau. Although the specific ionization of α radiation is far greater than that of β radiation, the count rates observed at the alpha plateau are significantly lower than those observed at the higher potentials. This is attributed to greater gas amplification at the higher potentials.

Experiment 18

The normalized curves intersect at 1,000 volts. When ^{133}Ba, ^{137}Cs, and ^{60}Co sealed sources are measured, ^{60}Co shows the earliest and lowest count rates, and ^{133}Ba shows the latest and highest count rates. These observations demonstrate that the pulses from the higher-energy photons require less amplification to trigger a count, and the higher-energy photons are more likely to escape from the detector without generating a luminescent photon to initiate the sequence of events leading to a count.

Experiment 19

The data shown in Figure 33 were collected using a 1 × 1 inch NaI(Tl) detector connected to a 100-channel analyzer set to count 30 seconds per

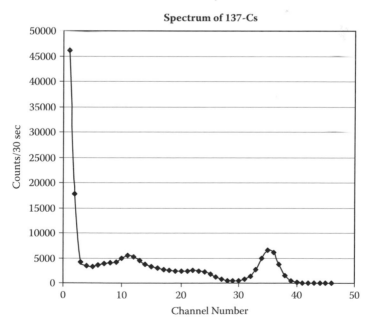

Figure 33 Spectrum of ^{137}Cs.

channel. These data show the photopeak centered at channel 35 and some of the other characteristics of a gamma spectrum. (See Bryan, 2009, Fig. 6.10, p. 108.)

Experiment 20

Spectral features will vary depending on the sources, and their activities. Typical sources and observed features (in addition to the photopeaks) include:

^{22}Na: Annihilation peak, sum peak, Compton continua and edges
^{137}Cs: Compton continuum and edge, back scatter peak, Ba x-ray peak
^{57}Co: Iodine escape peak
^{60}Co: Compton continua and edges, back scatter peak
^{65}Zn: Annihilation peak, back scatter peak, Compton continuum and edge
^{54}Mn: Compton continuum and edge

Experiment 21

The efficiencies for ^{3}H and ^{14}C will depend upon the instrumentation and the scintillation cocktail used. This experiment, however, will clearly demonstrate the quenching effect of the nitromethane and the dependency of the scintillation process on the aromatic solvent.

Experiment 22

This experiment demonstrates the need to use stable carriers for the successful isolation of ^{210}Pb and ^{210}Bi from the equilibrium system. The decay and ingrowth of ^{210}Bi in the bismuth and lead fractions becomes evident after approximately 3 weeks.

Experiment 23

The mobile phase can be prepared from the organic phase (upper layer) obtained after shaking 1 M hydrochloric acid with an equal volume of 1-butanol. Typical R_f values are 0.1 for the ^{210}Pb, 0.6 for the ^{210}Bi, and 0.8 for the ^{210} Po. Sometimes, the activity of the ^{210}Po cannot be detected. The presence of the ^{210}Bi is confirmed by its characteristic 5-day half-life.

Experiment 24

This experiment works well. Typical data for using RaDEF are as follows:

Trial	Activity	Trial	Activity	Trial	Activity	Trial	Activity
1	2,344	6	2,465	11	2,313	16	2,397
2	2,423	7	2,428	12	2,395	17	2,376
3	2,376	8	2,397	13	2,368	18	2,404
4	2,448	9	2,410	14	2,470	19	2,350
5	2,413	10	2,354	15	2,424	20	2,432

Mean = N = 2399; σ = √K11568_χομμA003ξ003.ρτφN = 49; s = √K11568_χομμA003ξ004.ρτφ[Σ(n_i − N)² ÷ (N − 1)] = 41

Trials within ±σ = 13. Trials beyond ±σ = 7.

Experiment 25

A cold run using distilled water is recommended to gain familiarity with and proficiency in physical manipulations involving small quantities of liquid reagents before using radioactive material.

Rarely are the samples shown to duplicates with a 90% probability.

Experiment 26

A 2- or 5-Ci PuBe neutron source or a 10-μg ^{252}Cf is suitable for this experiment, as well as for Experiments 24 and 25. The PuBe neutron source produces neutrons by an (α,n) reaction, i.e., $^4_2$He + $^9_5$Be → $^{12}_6$C + $^1_0$n, while the ^{252}Cf undergoes spontaneous fission to produce neutrons.

A = σφN(1 − e$^{-\lambda t}$), where A is the activity in dps at time zero, σ is the cross-section in barns for the ^{115}In (n,γ) ^{116m1}In reaction, φ is the neutron flux in cm^{-2} s^{-1}, N is the number of ^{115}In atoms in the indium disc, λ is the decay constant for ^{116m1}In in min^{-1}, and t is the duration of radioactivation in minutes. Indium is 4.23% ^{113}In and 95.77% ^{115}In. Cross-sections and half-lives are as follows:

^{113}In (n,γ) ^{114}In	σ = 4 b,	$t_{1/2}$ = 72 s
^{113}In (n,γ) ^{114m}In	σ = 8 b,	$t_{1/2}$ = 50 days
^{115}In (n,γ) ^{116}In	σ = 45 b,	$t_{1/2}$ = 14 s
^{115}In (n,γ) ^{116m1}In	σ = 145 b,	$t_{1/2}$ = 54 min
^{115}In (n,γ) ^{116m2}In	σ = 4 b,	$t_{1/2}$ = 2 s

Typical results for the half-life of ^{116m1}In range from 49 to 55 minutes. Results for the measurement of neutron flux produced by a-5 μg ^{252}Cf source ranged from 2 × 10³ to 3 × 10³ cm^{-2}s^{-1}.

Later results using a 2 Ci PuBe neutron source gave a half-life of 53 minutes and a neutron flux of 1.1 × 10⁴ cm^{-2}s^{-1}. Measurements made with a gold foil radioactivated for 4 days and counted periodically over a 5-day period showed a half-life of 66 hours and a neutron flux of 1.6 × 10⁴ cm^{-2}s^{-1}.

Experiment 27

The short half-life of the ^{56}Mn, 2½ hours, allows conventional waste disposal of the residues from this experiment after being allowed to undergo radioactive decay for a week. Care must be taken, however, to ensure the unknowns contain only aqueous $MnSO_4$ solutions.

Experiment 28

The short half-life of the ^{128}I, 25 minutes, allows conventional waste disposal of the residues from this experiment.

Based on the relative radioactivities of the two phases, approximately 90% of the ^{128}I remains in the organic layer.

Typical data for the decay of the ^{128}I are as follows:

Time	16.35	16.40	16.45	16.50	16.55	17.00
Activity, cpm	2,555	2,116	1,970	1,788	1,414	1,302

Experiment 29

Acetic anhydride is a corrosive agent. Skin contact must be avoided.

Before attempting to carry out this synthesis with radioactive material, a cold run is recommended to gain familiarity with and proficiency in the physical manipulations of small quantities of reagents.

The tracer ^{14}C-aetic anhydride should be diluted with carrier acetic anhydride to an activity of approximately 2½ μCi/ml.

Experiment 30

Sulfuric acid, acetic anhydride, and phosphorus pentachloride are corrosive agents. Skin contact must be avoided.

The synthesis of sulfanilamide is more involved than the synthesis of ^{14}C-acetyl salicylic acid (Experiment 29). Before attempting to carry out the synthesis of sulfanilamide with radioactive material, a cold run is recommended to gain familiarity with and proficiency in the physical manipulations of small quantities of reagents.

The carrier-tracer ^{35}S-sulfuric acid may be prepared by adding 50 μCi of ^{35}S-sulfuric acid in no more than 1 ml of water to 24 ml of concentrated sulfuric acid.

Experiment 31

Pocket dosimeters have been replaced by thermoluminescent dosimeter (TLD) devices for passive monitoring of laboratory personnel. However, first responders sometimes make use of the former. In addition, the pocket dosimeter can be used to reinforce the theoretical considerations of the electroscope used in Experiment 16. The 0 to 2 millirem (0 to 0.02 mSv) devices available from Supertech and from GeoData Systems are suitable for this experiment, as are left-over Cold War–era 0 to 200 millirem (0 to 2 mSv) CDV-138 Civil Defense training models.

Area and personnel monitoring with a handheld "pancake" Geiger-Müller detector as well as counting wet and dry wipes from the working area using a conventional, fixed location instrument can complete the monitoring exercise.

Experiment 32

Elements that work well as unknowns include V, Mn, As, Ag, In, I, La, Pr, Dy, and Au. Others that *can* work (given sufficient neutron flux and student tenacity) include Cu, Ga, Br, Rh, Pd, Sb, Sm, Eu, Ho, Er, Yb, W, Re, and Ir.

Appendix 4: some suppliers of radioactive materials and instrumentation for the detection and measurement of nuclear radiation

All Foils, Inc.
16100 Imperial Parkway
Cleveland, OH 44149

Telephone: 440-572-3845
Facsimile: 440-378-0161
Internet: www.allfoil.com

American 3B Scientific
2189 Flintstone Drive, Unit 0
Tucker, GA 30084

Telephone: 770-492-9111
Facsimile: 770-492-0111
Internet: www.A3bs.com

Canberra Industries, Inc.
800 Research Parkway
Meriden, CT 06450

Telephone: 203-238-2351
Facsimile: 203-235-1347
Internet: www.canberra.com

Daedalon Radioactivity
299 Atlantic Highway
Waldoboro, ME 04572

Telephone: 800-233-2490
Facsimile: 978-745-3065
Internet: www.daedalon.com

Fisher Science Education
4500 Turnbury Drive
Hanover Park, IL 60133

Telephone: 800-955-1177
Facsimile: 800-629-1166
Internet: www.fishersci.com

GeoData Systems Management Inc.
Berca, OH 44017-0366

Telephone: 440-888-4749

Images SI, Inc.
109 Woods of Arden Road
Staten Island, NY 10312

Telephone: 718-966-3690
Facsimile: 718-966-3695
Internet: www.imagesco.com

North American Technical Services
306 Industrial Park Road
Middletown, CT 06457

Telephone: 860-635-6820
Facsimile: 860-635-4962
Internet: www.nats-usa.com

Nuclear Technology Services, Inc.
635 Hembree Parkway
Rosewell, GA 30076

Telephone: 770-663-0711
Facsimile: 770-663-0547
Internet: ntsincorg.com

PASCO
10101 Foothills Boulevard
Roseville, CA 95747

Telephone: 916-786-3800
Internet: www.pasco.com

Sigma-Aldrich
Post Office Box 14508
Saint Louis, MO 63178

Telephone: 800-325-3010
Facsimile: 800-325-5052

Spectrum Technologies, Inc.
106 Union Valley Road
Oak Ridge, TN 37830

Telephone: 865-482-9837
Facsimile: 865-483-0473

Supertech
Post Office Box 186
Elkhart, IN 46515

Telephone: 547-264-4310
Facsimile: 574-264-9551

TEL Atomic, Inc.
Post Office Box 924
Jackson, MI 49204-0924

Telephone: 800-622-2866
Facsimile: 517-783-3213
Internet: www.telatomic.com

United Nuclear
Post Office Box 373
Laingsburg, MI 48848

Telephone: 517-651-5635
Internet: www.unitednuclear.com

Appendix 5: some suggested format for laboratory reports

Name of Experimenter   Name(s) of Collaborator(s)

Title of Experiment

Object of Experiment

Brief Description of Procedure

Tables and Graphs of Experimental Data

Calculations

Experimental Results

Theoretical Results

Conclusions

Index